Fourth Edition

Postharvest Technology of Horticultural Crops

Cooling and Storage

James F. Thompson
Cooperative Extension Specialist Emeritus
(UC ANR) in Agricultural Engineering
Department of Biological
and Agricultural Engineering
University of California, Davis

Irwin R. Donis-González
Associate Professor of
Cooperative Extension (UC ANR)
Department of Biological
and Agricultural Engineering
Co-Director Postharvest Technology Center
University of California, Davis

Tom Rumsey
Professor Emeritus
Department of Biological
and Agricultural Engineering
University of California, Davis

Farzaneh Khorsandi
Associate Professor of
Cooperative Extension (UC ANR)
Department of Biological and
Agricultural Engineering
University of California, Davis

Erin DiCaprio
Associate Professor of
Cooperative Extension (UC ANR)
in Community Food Safety
Department of Food Science and Technology
University of California, Davis

Volumes in the UC ANR Publication Series: *Postharvest Technology of Horticultural Crops*, Fourth Edition

The fourth edition of *Postharvest Technology of Horticultural Crops* expands significantly on the material covered in the third edition, published in 2002. The fourth edition will be published in ten volumes. Books are released as they are completed (rather than in the volume order listed below).

Volume titles and publication numbers

Volume 1: Preharvest, Harvest, and Packing Operations (21655)
Volume 2: Cooling and Storage (21656)
Volume 3: Atmospheric Environment (21657)
Volume 4: Quality Factors and Their Evaluation (21658)
Volume 5: Disease and Insect Control (21659)
Volume 6: Transportation and Destination Handling (21660)
Volume 7: Vegetable Crops, Herbs, and Flowers (21661)
Volume 8: Fruit Crops and Tree Nuts (21662)
Volume 9: Fresh-Cut and Processed Horticultural Products (21663)
Volume 10: Extension Methods and Capacity Building (21664)

All published volumes are available through the UC ANR Publications Catalog: anrcatalog.ucanr.edu

University of California
Agriculture and Natural Resources
Davis, California
Publication 21656

UNIVERSITY OF CALIFORNIA
Agriculture and Natural Resources

To order or obtain UC ANR publications and other products, visit the UC ANR online catalog at https://anrcatalog.ucanr.edu/ or phone 1-800-994-8849. Direct inquiries to

UC Agriculture and Natural Resources
Publishing
2801 Second Street
Davis, CA 95618

Telephone 1-800-994-8849
E-mail: anrcatalog@ucanr.edu

Fourth edition, 2023

Publication 21656

ISBN-13: 978-1-62711-059-4

Library of Congress Cataloging-in-Publication Data

Names: Thompson, James F. (Postharvest Specialist Emeritus), author. | Rumsey, Tom, author. | Khorsandi, Farzaneh, author. | Donis-González, Irwin R., author. | DiCaprio, Erin, author.

Title: Postharvest technology of horticultural crops : cooling and storage / James F. Thompson, Tom Rumsey, Farzaneh Khorsandi, Irwin R. Donis-González, Erin DiCaprio

Other titles: Publication (University of California (System). Division of Agriculture and Natural Resources) ; 21656.

Description: Fourth edition. | Davis, CA : University of California Agriculture and Natural Resources, 2023 | Series: Publication ; 21656 | Volume Two of the Fourth edition of Postharvest Technology for Horticultural Crops. | Includes bibliographical references and index. | Summary: "Cooling and Storage, a volume in the fourth edition of the Postharvest Technology for Horticultural Crops series, covers crucial issues in the commercial cooling of fruit, vegetables and flowers. The topics it discusses include proper temperature management practices; commercially used methods for cooling fruit, vegetables, and flowers; the design of cold-storage facilities; worker safety and food safety for cooling and storage systems; and the effects of air temperature and humidity on postharvest quality, as well as temperature and humidity measurement methods"-- Provided by publisher.

Identifiers: LCCN 2023035594 | ISBN 9781627110594 (paperback)

Subjects: LCSH: Horticultural crops--Postharvest technology. | Horticultural crops--Postharvest physiology. | Plant products--Postharvest technology.

Classification: LCC SB319.7 .T48 2023 | DDC 635--dc23/eng/20230823

LC record available at https://lccn.loc.gov/2023035594

UC PEER REVIEWED This publication has been anonymously peer reviewed for technical accuracy by University of California scientists and other qualified professionals. This review process was managed by UC ANR Associate Editors for Postharvest Technology and Biology Florence Zakharov and Margaret Lloyd.

Art Direction: Sandra Osterman | Design: Brennan Collins

11/23-LC/BC

Statement on pesticide mentions in *Postharvest Technology of Horticultural Crops: Cooling and Storage*

Discussions in this book of pesticides (including fungicides, insecticides, and sanitizing agents) are not to be considered recommendations for a particular purpose, as conditions vary. The authors provide basic information about pesticides available or in development at the time of the book's publication. Please consult an expert before applying any pesticide.

Any usage of pesticides described herein must be based on the product label and must comply with all applicable local, state, and federal rules regarding usage rates, disposal, and safety.

Pesticide application in commercial settings is highly regulated in California. Please consult with your county agricultural commissioner or regulatory representative before pesticide application.

The University of California does not accept liability for any and all claims, losses, costs, damages, expenses (including attorneys' fees), and personal injuries (up to and including death) due to the application of the methods in this book.

Contents

Series preface

About one-third of fresh produce harvested worldwide is lost at various points in the distribution system between production and consumption. While it is impossible and uneconomical to eliminate these losses completely, it is possible to reduce them by at least half and increase food availability for the rising world population. Minimizing postharvest losses is more sustainable and environmentally sound than compensating for the losses by increasing the land and resources devoted to food production. The strategies for reducing postharvest losses include selecting genotypes with improved postharvest quality and longer postharvest life, using production methods that optimize postharvest quality, and using postharvest handling procedures to maintain produce quality.

Great advances have been made in postharvest technology through cooperative research by plant scientists, engineers, and food scientists. This interdisciplinary research, coupled with effective extension programs, needs to continue—to further improve the quality, flavor, nutritional value, and safety of perishable produce.

The goal of the technical editors and the authors of this *Postharvest Technology of Horticultural Crops* series is to describe the current scientific understanding of quantitative and qualitative losses in fresh-market vegetables, fruit, tree nuts, and flowers. The emphasis is on commercially available technologies that have been placed into practice to minimize losses and maintain food safety. Our perspective is based on the postharvest technology used in California and the United States. However, the principles discussed are applicable to fresh horticultural crops produced anywhere in the world. The information will be useful to people who want to learn about or are involved in growing, harvesting, packaging, transporting, or marketing perishable produce.

The *Postharvest Technology of Horticultural Crops* series is the outcome of a syllabus that was developed for a short course initiated in 1979 and offered annually since then in two modes: as a regular UC Davis course (Plant Science 196) for advanced undergraduate and graduate students and as a short course organized by the UC Davis Postharvest Center for participants who are not current UC Davis students. The latter group includes research and extension workers, consultants, quality and safety inspectors, packers, storage operators, transportation company representatives, and other professionals involved in postharvest handling of fresh horticultural perishables.

The first (1985), second (1992), and third (2002) editions of this series were published as single books and were well received, widely distributed, and used throughout the world. The amount of information on postharvest handling has continued to expand, and so this fourth edition has been converted into a series of ten books, each covering a separate area of postharvest technology.

We are indebted to our former colleagues Robert F. Kasmire, F. Gordon Mitchell, and Noel F. Sommer, who contributed greatly to the previous editions. Dr. Adel A. Kader initiated this fourth edition but was unable to see the conclusion of the publication because of his untimely death. We dedicate this series to him and have kept him as the lead technical editor on the series. He devoted his professional career to the cause of improving the postharvest handling of perishables. His enthusiasm, creativity, and unflagging commitment to his profession were an inspiration to us and to postharvest professionals around the world. We are indebted to Dr. Kader for his passion for bringing more and better-quality fruit and vegetables to the world's consumers.

On behalf of the authors, we also wish to thank all those who assisted in developing this series, especially Mary Reed, former office manager of the Postharvest Center. We also thank the staff of UC Agriculture and Natural Resources Communication Services who participated in the production of this series.

James F. Thompson, Mikal Saltveit
Technical Editors

Preface to *Cooling and Storage*

The first chapter of this volume is the fourth revision of Coolers for Fruits and Vegetables by Rene Guillou, published in 1960 (Guillou 1960). It was first revised in 1972 by F. G. Mitchell, Rene Guillou, and R. A. Parsons and retitled Commercial Cooling of Fruits and Vegetables (Mitchell et al. 1972). In 1998, J. F. Thompson, F. G. Mitchell, T. R. Rumsey, R. F. Kasmire, and C. H. Crisosto made the third revision, titled Commercial Cooling of Fruits, Vegetables, and Flowers (Thompson et al. 1998). It was slightly revised in 2002 (Thompson et al. 2002).

The chapter describes both proper temperature management practices for perishable commodities and the commercially used methods for cooling fruit, vegetables, and cut flowers. It is written for a person who is initially investigating produce cooling, a professional designer who needs design details, and an operator who wants a better understanding of practical operation guidelines. The chapter contains a complete discussion of design for forced-air coolers, hydrocoolers, and vacuum coolers—the most commonly used cooling methods that people with a good background in industrial refrigeration can design. There is little discussion of refrigeration equipment design because many other good publications deal with that subject, including Stoecker (1988–1995) and Dossat and (Horan 2002), as well as publications by the American Society of Heating, Refrigerating, and Air-Conditioning Engineers; the Refrigerating Engineers and Technicians Association; and the International Institute of Ammonia Refrigeration.

The second chapter is an overview of cold storage for perishables and is an update of the chapter on storage systems in Kader (2002). It describes the unique issues associated with designing a cold storage for perishables but does not contain a detailed description of refrigeration load calculations or equipment selection. The books listed above cover the details of these issues.

Worker safety and food safety for cooling and storage systems have become important issues for the industry, and they are discussed in chapters 3 and 4. The volume concludes with chapter 5, which describes the effects of air temperature and humidity on postharvest quality and temperature and humidity measurement methods. This chapter is based on "Psychrometrics and Perishable Commodities" in Kader (2002).

References

Dossat, R. J., T. J. Horan. 2002. Principles of refrigeration. 5th ed. Upper Saddle River, NJ: Prentice Hall.

Guillou, R. 1960. Coolers for fruits and vegetables. Bulletin of the California Agricultural Experiment Station 773.

Kader, A. A., ed. 2002. Postharvest technology of horticultural crops. Oakland: UC Agriculture and Natural Resources Publication 3311.

Mitchell, F. G., R. Guillou, and R. A. Parsons. 1972. Commercial cooling of fruits and vegetables. UC Agricultural Experiment Station Manual 43.

Stoecker, W. F. 1988–1995. Industrial refrigeration. Vols. 1–2. Troy, NY: Business News Publishing.

Thompson, J. F., F. G. Mitchell, T. R. Rumsey, R. F. Kasmire, and C. H. Crisosto. 1998. Commercial cooling of fruits, vegetables, and flowers. Oakland: UC Agriculture and Natural Resources Publication 21567.

Thompson, J. F., F. G. Mitchell, T. R. Rumsey, R. F. Kasmire, and C. H. Crisosto. 2002. Commercial cooling of fruits, vegetables, and flowers. Oakland: UC Agriculture and Natural Resources Publication 21567.

Chapter 1
Initial cooling

James F. Thompson and Tom Rumsey

The postharvest life of perishable fruit, vegetables, and flowers is affected by many factors, including produce respiration, compositional changes, water loss, exposure to high or low temperatures, exposure to ethylene, mechanical damage, and decay. Produce temperature directly or indirectly affects all these factors and is the major factor controlling postharvest life. Cooling perishables after harvest is the key to providing consumers with good-quality fruit, vegetables, and flowers.

The duration of postharvest life is usually based on appearance. Shriveled fruit, wilted vegetables, and visible decay cause produce to be rejected at the market. Consumer acceptance is also based on flavor, which begins to decline before produce appears damaged. Generally, the period of acceptable taste lasts about two-thirds as long as the period of acceptable appearance, and nutrition declines even earlier than taste.

The potential postharvest life of fresh fruit, vegetables, and flowers is quite variable—lasting from a few days for many flowers to many months for items like apples, potatoes, and winter squashes; table 1.1 shows the typical postharvest life of a few commodities. Growing conditions, the cultivar raised, and specific handling and storage practices also affect the storage life of any individual lot of produce. For some items, postharvest life can be extended beyond the times listed in table 1.1 with modified or controlled atmospheres, chemical treatments, or edible coatings.

Causes of postharvest losses

Fruit, vegetables, and flowers are living organisms that undergo all the physiological and pathological processes associated with plants. Good postharvest quality is obtained by controlling the causes of produce quality loss and damage (table 1.2). If produce is eaten or used within a few hours after picking, there is little need to manage the postharvest environment of

Table 1.1. Postharvest life of selected fresh produce based on appearance and assuming items are held continuously at their minimum storage temperature

Postharvest life	Vegetables	Fruits and melons
<2 weeks	artichokes, beans, cucumbers, sweet corn, mushrooms, peas, salad greens, summer squash	most berries, cantaloupes, cherries, papayas, strawberries
2–4 weeks	broccoli, eggplant, lettuce, green onions, peppers, radishes, rhubarb	avocados, mandarins, mangoes, nectarines, peaches, pineapples, watermelons
4–8 weeks	cauliflower, celery	grapefruit, pomegranates
8–16 weeks	jicamas	limes, oranges
>16 weeks	carrots, garlic, sweet potatoes, turnips	kiwifruit, many apple varieties

Table 1.2. Environmental factors that influence postharvest deterioration mechanisms of perishables

Deterioration mechanism	Environmental factors						
	Temperature	Humidity (vapor pressure difference)	Ethylene	CO_2 and O_2 levels	Chemical treatments, surface coatings	Packaging (other than modified atmosphere packaging)	Light
respiration	yes	no	yes	yes	yes	no	no
compositional changes	yes	no	yes	yes	yes	no	yes
water loss	yes	yes	no	no	yes	no	no
decay	yes	yes	yes	yes	yes	no	no
chilling, freezing, and high-temperature damage	yes	no	no	no	no	no	no
mechanical damage	yes	no	yes	no	no	yes	no

produce. But the short harvest season of most perishables cannot meet consumer demand for long-term availability of perishables. Also, production areas are usually distant from markets, requiring a few days to several weeks for transport from farm to market. Maintaining the lowest possible produce temperature is the most effective way to extend postharvest life and reduce postharvest loss. Understanding the relationship between produce temperature and deterioration mechanisms is essential to reducing postharvest loss.

Respiration

Fresh flowers, vegetables, and fruit are alive. After harvest they use the process of respiration to produce the energy they need to continue living. Respiration is a complex series of enzyme-driven chemical reactions (fig. 1.1). Oxygen from the air reacts with starches, sugars, and other components stored in plant tissues to produce adenosine triphosphate (an energy transport molecule), carbon dioxide, water, and heat.

Respiration is the greatest determinant of how quickly fresh produce deteriorates. If oxygen absorption is prevented, normal respiration cannot proceed, and fermentation quickly destroys the produce. Also, heat released by respiration can contribute to produce warming during storage and transportation.

Respiration is regulated by the catalytic action of protein molecules called enzymes. Enzyme activity is very sensitive to temperature, and respiration increases by a factor of about 2.7 for each 10°C (18°F) of increased temperature (fig. 1.2) up to temperatures of about 25° to 30°C (77–86°F). Above this temperature range,

$$C_6H_{12}O_6 + 6O_2 \xrightarrow{\text{Enzymes}} 6CO_2 + 6H_2O + ATP + \text{heat}\ (686\ \text{kcal-mole}^{-1})$$

Figure 1.1. Simplified description of plant respiration.

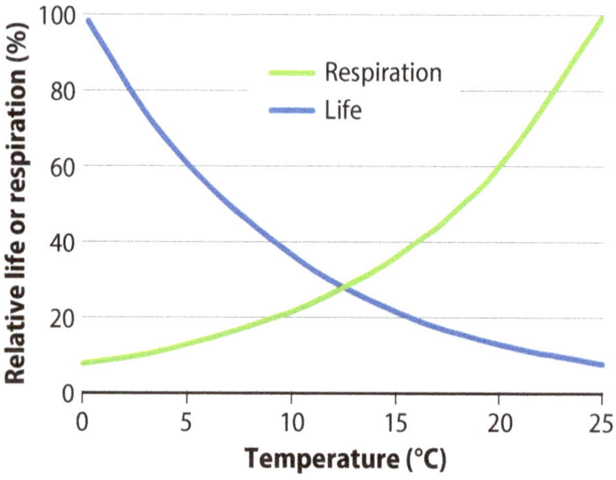

Figure 1.2. Effect of produce temperature on relative respiration rate of fruit and vegetables and relative postharvest life.

respiration rates increase more slowly, and even higher temperatures damage or kill produce. Holding non–chilling sensitive produce items at 25°C (77°F) rather than their recommended storage temperature of 0° (32°F) may reduce postharvest life by 95 percent (see fig. 1.2).

Appendices A-1 and A-2 list the typical postharvest life of a variety of fruit, vegetables, and flowers. These approximations assume each produce item is held at its recommended storage temperature. If an item is held at a higher-than-recommended temperature, its storage life will decrease at a rate proportional to its increased respiration rate, assuming that its quality is not prematurely affected by decay, water loss, or some other factor. For example, if a perishable item is held for 1 hour at 10°C (18°F) above its recommended storage temperature, it will lose 2.7 hours of postharvest life. If the item is held for 1 hour at 20°C (36°F) above its recommended storage temperature, it will lose 7.3 hours of postharvest life (2.7 × 2.7).

Compositional changes

Some perishable commodities undergo beneficial changes after harvest, and their flavor and appearance improve. Sometimes this occurs in low-temperature storage, as with the improved flavor development of some apple varieties and the more uniform ripening of pears. Other fruit, such as bananas, avocados, and tomatoes, are held near room temperature and at high relative humidity, and may be exposed to ethylene to cause beneficial softening, color change, or increased sugar levels. Stone fruit are less susceptible to chilling injury if they are held near room temperature for about a day immediately after harvest.

With many other commodities, compositional changes are detrimental because they soften and become more subject to mechanical damage. Nearly all commodities lose nutritional value and flavor over time, and the rate of loss increases with increasing temperature. Generally, environmental conditions that slow respiration also slow the loss of nutrients and flavor. Exposure to light can also cause detrimental compositional changes in a few commodities. For example, Belgian endive gains an undesirable green color, and potatoes produce a toxic substance called solanine.

Water loss

Wilting and shriveling seriously damage the appearance of produce and reduce its consumer appeal and market value. Some perishables, particularly leafy vegetables, appear shriveled or wilted after water loss equaling only a small percentage of their weight at harvest (table 1.3). Stem shrivel can occur in grapes (Crisosto et al. 1994) and cherries with weight loss of less than 2 percent. Other produce items do not show symptoms until higher levels of weight loss, and their postharvest life may be more limited by decay, physiological damage, or loss of taste (Robinson et al. 1975).

Table 1.3. Water loss at which commodities are unsalable

Commodity	Maximum weight loss (% fresh weight)	Reason for loss
spinach	3	wilting
broccoli	4	taste, wilting
turnips with leaves	4	wilting
tomatoes	4	shrivel
leaf lettuce	3–5	wilting, decay
grapes	5	berry shrivel
pears	6	shrivel
cabbage	6	shrivel
apples	7	shrivel
watercress	7	wilting
persimmons	7	shrivel
carrots	8	wilting
Brussels sprouts	8	wilting, rot, yellowing
green peppers	8	shrivel
peaches	11	shrivel
winter squash	15	hollowneck

Sources: Hardenburg et al. 1986; Hruschka 1977; Nelson 1985; Robinson et al. 1975.

Reduced weight caused by water loss is a direct marketing loss. A weight loss of 5 percent means 1 pound less of salable produce in each 20-pound package. Water loss may cause additional losses. For example, wilted leafy produce may require trimming to make it marketable; grape berries may separate from the cluster if their pedicles are dry; and shriveled fruit must be discarded before sale. Water loss also weakens plant cells, making cells more susceptible to pathogens, which may result in greater ethylene production and increased chlorophyll loss, causing yellowing.

Water escapes from produce in the form of water vapor. Plants are composed of cells loosely bound together, with a considerable number of interconnecting intercellular spaces (fig. 1.3). Water from the cells vaporizes into the intercellular spaces and maintains a nearly saturated atmosphere within a produce item. Water vapor moves to the outside atmosphere through openings in the surface of the produce: lenticels, stomata, stem scars, injured areas, or directly through the cuticle (fig. 1.4). The water loss from a fruit or vegetable is driven by the vapor pressure difference between the item and the air surrounding it (Becker and Fricke 1996). Moisture loss is also influenced by an item's ratio of surface area to volume, as well as the speed of air flowing past the produce item.

The total effect of produce water-loss mechanisms, surface-to-volume ratio, and air speed is described by a measurement called the transpiration coefficient, which is the milligram-per-second moisture-loss rate per kilogram of produce, per megapascal of vapor-pressure difference between the produce item and the surrounding air. (Surface area of many commodities is difficult to measure, so transpiration coefficient is based on the mass of a perishable rather than its surface-to-volume ration.) Transpiration coefficients vary widely among different produce items (table 1.4). Notice that items with a spherical shape and a correspondingly low surface-to-volume ratio have low transpiration coefficients. Leafy vegetables with very high surface-to-volume ratios have transpiration coefficients one hundred times higher than many spherical produce items.

The rate of produce moisture loss is related to a produce item's transpiration coefficient, the temperature difference between the item and its surrounding air, and the humidity of the air around the item. (Calculating moisture loss of packaged produce is a complex task [Tanner et al. 2002].) The last two factors are combined in vapor-pressure difference (VPD), which is the vapor pressure in the interior of a commodity minus the vapor pressure of the air surrounding the commodity. High VPD causes rapid water loss.

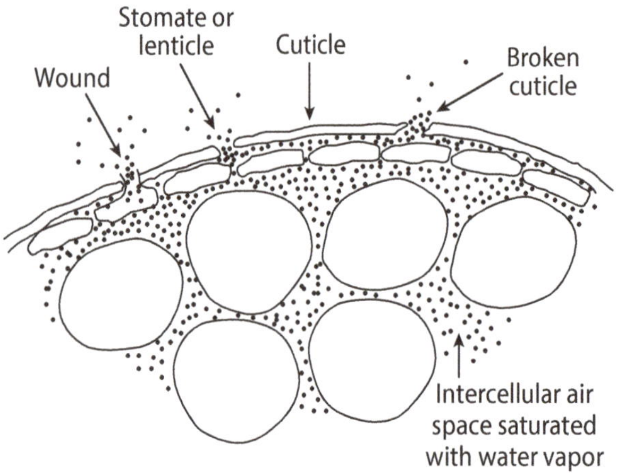

Figure 1.3. Primary sources of water loss in fresh produce.

Figure 1.4. Air injected into a pear is transmitted through intercellular spaces to lenticels on the surface of the fruit.

Vapor pressure is calculated with the following equation:

$$Vp = (w \cdot Pa)/0.622$$

where

Vp = vapor pressure (in same units as Pa)

w = humidity ratio (kg of water vapor per kg of dry air, usually calculated using air temperature and relative humidity)

Pa = atmospheric pressure (megapascals or inches of mercury), and

0.622 = ratio of the molecular weight of water divided by the molecular weight of air.

Atmospheric pressure is the same for both the produce item and the air, so VPD is directly related to the difference in humidity ratio (mass of water vapor per mass of dry air). At ambient temperature, the air in a fruit or vegetable is at 98 to 99 percent relative humidity, nearly saturated (Chau et al. 1987), but the air outside is usually at a lower humidity ratio. Produce needs to be kept in a cold environment to slow respiration, but in a refrigerated room, cold air has a low humidity ratio because it cannot hold as much water vapor as warm air. Warm produce in a cold room is exposed to a large VPD and should be cooled quickly to minimize VPD and water loss.

After cooling, it is important to keep a high relative humidity around the produce. Most perishables should be held in cold-storage conditions near 95 percent relative humidity, although a few that lose moisture slowly, such as citrus fruit, are held near 90 percent relative humidity to reduce decay growth. High humidity around the produce can be attained by keeping the humidity high in the refrigerated space or by covering it with a moisture barrier, like a plastic bag. Some produce items stored for long periods are packed in plastic containers that do not absorb water from the produce and are not weakened by high relative humidity.

Air flowing past a produce item increases its transpiration coefficient (Becker and Fricke 1996) and speeds moisture loss, so forced-air cooling may not be appropriate for some wilting-sensitive items. Even items suited to forced-air cooling should be protected from unnecessary airflow after cooling. Perishable produce can lose water rapidly when exposed to windy conditions or during transport on an open vehicle.

Table 1.4. Transpiration coefficients for selected fruit and vegetables

Commodity	Transpiration coefficient ($mg \cdot kg^{-1} \cdot sec^{-1} \cdot MPa^{-1}$)	Range of coefficients reported in literature ($mg \cdot kg^{-1} \cdot sec^{-1} \cdot MPa^{-1}$)
apples	42	16–100
potatoes	44	2–171
onions	60	13–123
pears	69	10–144
grapefruit	81	29–167
oranges	117	25–227
grapes	123	21–254
plums	136	110–221
tomatoes	140	71–365
lemons	186	139–229
cabbage	223	40–667
rutabagas	469	—
peaches	572	142–2,089
leeks	790	530–1,042
carrots	1,207	106–3,250
celery	1,760	104–3,313
parsnips	1,939	1,094–2,771
Brussels sprouts	6,150	3,250–9,770
lettuce	7,400	680–8,750

Source: Adapted from Sastry et al. 1978.
Note: "—" indicates that data are not available.

Water loss is cumulative and, like temperature, must be managed throughout the handling chain. Figure 1.5 depicts the influence of management variables on water loss in grapes. Although more severe handling practices (heavy line in the figure) result in greater weight loss, the treatments

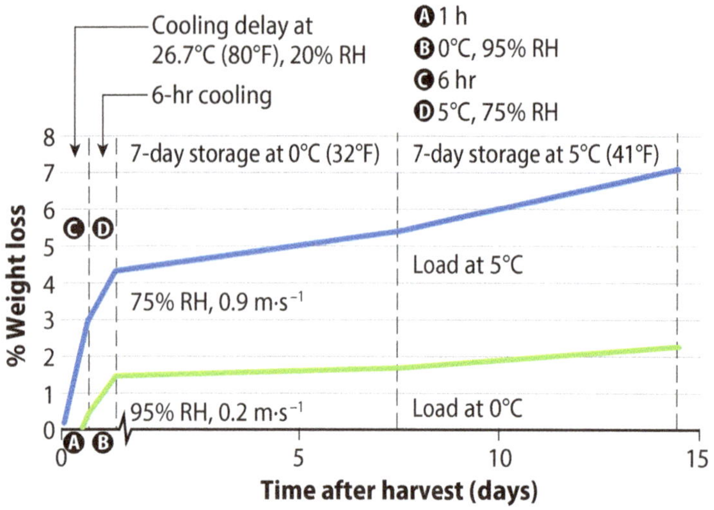

Figure 1.5. Table-grape weight loss resulting from two different cooling and handling regimes. The lower line represents nearly optimum conditions and the upper line represents the effects of poor handling.

shown are less extreme than those sometimes encountered in commercial practice.

The following observations about relative humidity, air temperature, and air speed can be made from the example in figure 1.5:

- Each hour of exposure to warm, dry air results in more than twice as much water loss as holding the grapes in high-humidity cold storage for 1 week.
- Use of colder and more-humid air during cooling reduces water loss slightly and cools produce to a lower, more desirable temperature for storage or shipment.
- Proper storage conditions (cooling grapes to near storage temperature, reducing air speed, and humidifying storage air) reduce water loss to 0.2 percent of the grapes' weight in 3 or 4 days.
- Grapes cooled to 0°C (32°F) lose less weight in transport at 4°C (40°F) than grapes cooled to 4°C (40°F).

Ethylene

Ethylene gas (C_2H_4) is produced by most plant materials, but many types of tropical fruit have especially high production rates. Ethylene can have beneficial or detrimental effects during postharvest handling. It induces beneficial ripening in fruit such as bananas, avocados, and European pears, and is used to degreen citrus fruit. However, ethylene concentrations near 1 part per million can cause damage such as russet spotting in iceberg lettuce, petal drop in carnations, and green color loss in cucumbers and leafy vegetables. Damage often occurs after about 24 hours of exposure to harmful concentrations, and exposures are cumulative. For most commodities, the maximum effect occurs when produce temperatures range between 16° and 21°C (60° and 70°F). A few produce items are even more sensitive to damage. For example, exposure to an ethylene concentration of only 5 to 10 parts per billion can cause premature softening in kiwifruit in cold storage. Low produce temperature reduces ethylene production, produce sensitivity to ethylene, and the rate of damage development. Rapid and complete cooling and good temperature management help to reduce ethylene-caused damage.

Ethylene concentrations can be reduced by eliminating internal combustion engines from cold rooms, periodically venting the storage room with outside air, using ethylene-adsorbing materials or scrubbers, and venting ripening rooms to the outside before opening them. Ethylene damage to some produce items can also be minimized with chemical treatments that make these items less sensitive to ethylene.

Decay

Decay-causing fungi and bacteria that attack fresh produce are also living organisms and are major contributors to postharvest loss. Some decay organisms can directly penetrate healthy tissue; many others enter only after produce has been weakened or wounded. Temperatures favorable for maximum growth and development of rot organisms often coincide with temperatures encountered during harvest. Rapid handling and cooling minimize decay development. The inhibiting effect of low temperatures varies with the organism. Postharvest pathogens such as *Botrytis cinerea* (gray mold), *Penicillium expansum* (blue mold), *Alternaria alternata*, *Cladosporium herbarum*, and *Monilinia fructicula* (brown rot of stone fruit) continue to grow, although very slowly, below 0°C (32°F). Other fungal rot organisms, such as *Rhizopus stolonifera* and *Aspergillus niger*, do not grow at temperatures below 5°C (41°F). Holding perishable commodities at low temperatures is an important tool in controlling postharvest decay.

Chilling, freezing, and high-temperature damage

Near-freezing temperatures that are ideal for many commodities may damage others, especially fruit, vegetables, and flowers of tropical and subtropical origin. This damage is called chilling injury. Symptoms of chilling injury include tissue darkening and drying, surface pitting, failure to ripen normally, off flavors, and increased susceptibility to decay. Some commodities show chilling injury only after prolonged storage, whereas others may be injured by brief exposures to low temperatures. Severity of chilling injury increases as exposure time increases and as the magnitude of temperature drop below the chilling threshold temperature increases. Chilling injury is cumulative.

Produce items not sensitive to chilling injury are usually cooled to near their freezing temperature. Although water freezes at 0°C (32°F), fruit and some vegetables freeze at temperatures several degrees lower because of their high soluble-solids content. The highest freezing temperatures for selected perishable produce items are listed in appendix A-1.

Mechanical damage

Bruises, cuts, abrasions, and other injuries cause serious produce loss during distribution and marketing, and usually speed deterioration from other causes. Mechanical damage to plant tissues accelerates respiration and often induces ethylene production. Mechanical damage also disrupts a perishable item's natural moisture barrier, which accelerates water loss, and damaged areas can become entry points for decay organisms.

Most produce items are more susceptible to compression and impact damage when they are cold than when they are warm. For best quality, produce should be packed soon after harvest and then cooled to storage temperature. If there are significant delays between picking and packing, quality can be preserved by cooling produce to an intermediate temperature before packing, perhaps 5° to 10°C (41–50°F), and finishing the cooling process after packing.

In contrast to impact and compression bruising, vibration damage is more serious when fruit is warm. This type of damage occurs during transit, when pieces of fruit vibrate or rub against each other or packing materials. Perishable produce should be shipped at its coolest possible temperature. Vibration damage during transport can be reduced by using vehicles equipped with air-ride suspension systems; packing fruit in polyethylene bags; not stowing produce sensitive to vibration damage in the rear of a vehicle, where vibration levels are greatest (Hinsch et al. 1993); and immobilizing produce with tight-fill packing techniques (Mitchell et al. 1968).

Initial cooling: Part of a temperature management system

For maximum postharvest life, most produce must be kept as cold as possible for as long as possible. A few types of fruit are ripened or conditioned at room temperature before initial cooling, for better flavor or texture. This period can last for 1 to several days. After this, the fruit must be cooled to its lowest possible storage temperature.

Temperature management begins as soon as produce is picked. Aboveground fruit and vegetables are generally close to the ambient temperature. Shaded produce will be slightly cooler than the ambient air temperature, and produce in direct sunlight can be significantly warmer than the surrounding air (Bergqvist 2001). Harvest while air temperatures are coolest, usually in the morning. Sometimes perishables are harvested late at night, in the hours before sunrise, with the aid of lights. If the produce must be stored temporarily before cooling, keep it in a shaded area. Cover the produce boxes with lids if they are left temporarily in the open sun (fig. 1.6). Avoid extended temporary storage at harvest temperatures by sending trucks to the cooler on a schedule, even if they are not filled. Tests with a few representative commodities have shown that short delays between picking and the start of cooling are commercially acceptable (table 1.5). However, reducing the time between picking and the start of cooling still improves postharvest life.

All subsequent handling operations after cooling—including storage, transportation, and marketing—should be designed and operated to keep produce at its lowest possible temperature.

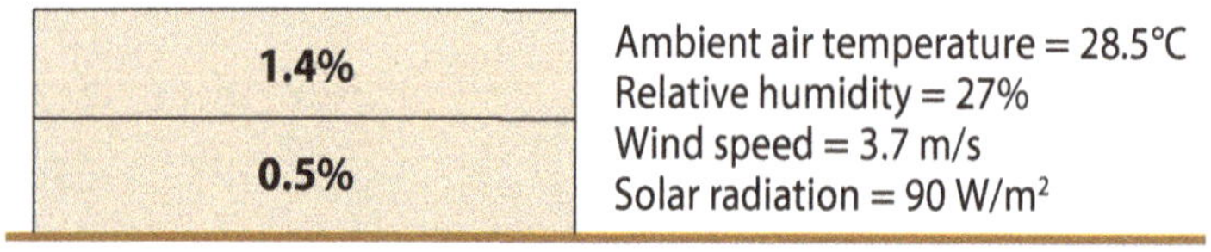

Figure 1.6. Average weight loss of single-layer cauliflower packed in open-top fiberboard boxes and placed in the open sun for 3 hours. The cauliflower in the top box is exposed to direct sunlight while the cauliflower in the bottom box is protected from the sun by the box above it.

Table 1.5. Acceptable delay between harvest and the start of cooling for selected fruit and vegetables

Acceptable delay (hr)	Fruit and melons	Vegetables
2	strawberries, grapes if >30°C	—
4	apricots, cantaloupes	asparagus, mushrooms, leafy green vegetables, sweet corn
8	avocados, grapes if <20°C, kiwifruit, mandarins, peaches, nectarines	artichokes, bell peppers, cauliflower, carrots, summer squash, cucumbers, eggplant, green beans, green onions
16	honeydews, oranges, pears, persimmons, plums	garlic, dry onions, peppers, potatoes, sweet potatoes, tomatoes
24	apples, grapefruit, lemons	—

Note: "—" indicates that table cells are left intentionally blank.

Regularly monitoring and recording produce temperature throughout the handling chain are key to guaranteeing that consumers receive consistent and high-quality produce.

Maintaining low temperatures may not be possible at all steps in the handling chain, but tests have shown that keeping produce at its optimum storage temperature for only a portion of the distribution time is better than no refrigeration at all. Even produce undergoing repeated cycles of cooling and warming deteriorates at a slower rate than produce that has not been cooled at all. Tests with cherries (fig. 1.7) indicate that higher handling temperatures result in much greater loss (from physiological deterioration, fruit rot, and desiccation), even within a 2- or 3-day marketing period. In some cases, reducing the time produce spends in the handling chain can compensate for less-than-ideal produce temperatures. For example, air transport is used for some high-value, highly perishable items sold to overseas markets, even though airplane holds are kept near room temperature. Rapid marketing, however, only partially offsets the value of proper cold-chain handling.

Room cooling

Exposing produce containers to cold air in a refrigerated space is a common method (Guillou 1960; Sainsbury 1961) for cooling cut flowers before packing, initially cooling apples and pears in a controlled-atmosphere storage room,

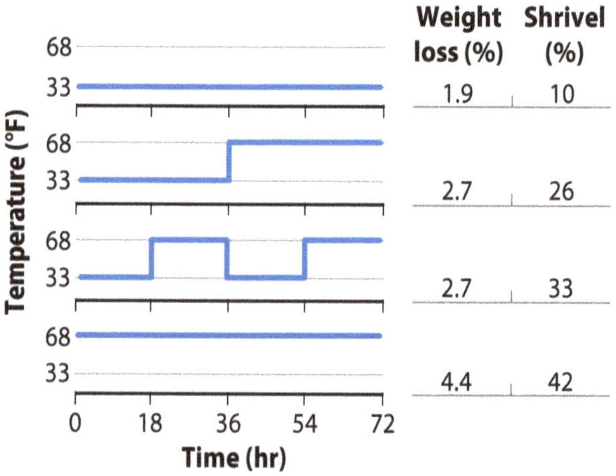

Figure 1.7. Effect of holding cherries at four temperature regimes.

and initially cooling potatoes, sweet potatoes, and citrus fruit. It is well suited for finish-cooling of produce that has been cooled to within a few degrees of optimal temperature after packing, such as hydrocooled cherries packed in plastic liners and fresh-cut vegetables.

A simple room-cooling system discharges cold air into a storage room horizontally, just below the ceiling. The air sweeps past the ceiling and then flows past produce boxes stacked on pallets (fig. 1.8). In cold rooms more than 15 meters (50 ft) long, ceiling ducts may be needed to ensure that cooled air reaches all parts of the room (fig. 1.9). Air flowing out of the ceiling ducts or a plenum produces good flow past the end pallets and pallets near the floor. Other methods for good airflow distribution are described in the storage section that follows.

The fastest possible room cooling requires a fan capacity of 0.3 $m^3 \cdot min^{-1} \cdot t^{-1}$ (100 cfm/ton), based on the maximum produce capacity of the cold room. The air-handling system must be designed to distribute air uniformly throughout the cold room. When initial cooling is completed, moisture loss can be minimized by reducing airflow to the lowest level that will keep the produce uniformly cool in all locations in the storage room. Typically, airflow can be reduced to 20 to 40 percent of that needed for initial cooling.

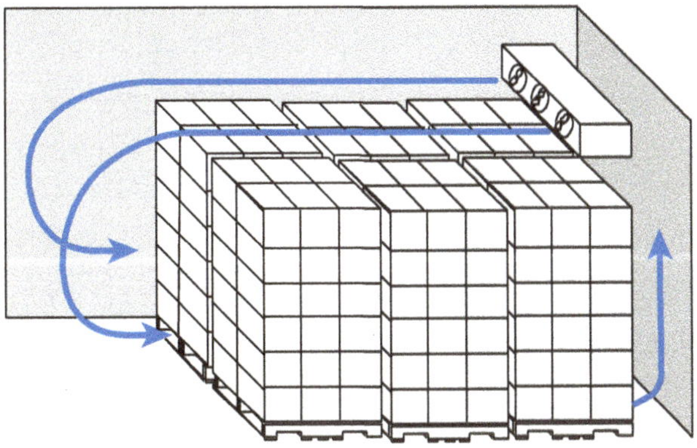

Figure 1.8. Airflow pattern in a room cooler with unit evaporators.

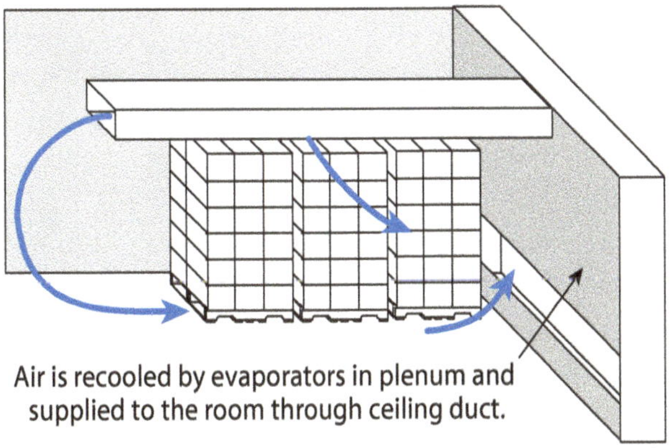

Figure 1.9. Airflow pattern in a room cooler with a slotted ceiling duct.

The fastest possible room cooling also requires that boxes be stacked in a way that allows air to flow past them (Guillou 1963; Mitchell et al. 1971). Spaced stacking of boxes on a pallet allows air to flow past several sides of each box, but the boxes must be squeezed together later for long-term storage or shipping. Well-vented containers, with vents that align between stacked boxes, cool four times faster than closed containers (table 1.6). Vent shape, location, and other design constraints are the same for room cooling as for forced-air cooling, and will be described later.

Use the lowest possible air temperature in the cold room. This is usually 0.5° to 1°C (1–2°F) below the freezing or chilling temperature of the perishable.

Room cooling should be monitored by measuring the air temperature in several locations, to locate warm or cold areas in the room. Temperature-sensing units should not be placed on exterior walls, where heat from outside will cause erroneously high readings. Warm areas are usually a result of poor airflow. Change the air-distribution system or the pallet placement to increase airflow to warm areas.

Table 1.6. Factors affecting room-cooling time of pallet loads of tight-fill packed plums

Vent area (% of side wall area)	Packing spacing in cm (in.)	Stacking pattern	Longest cooling time (hr)
0	0 (0)	vents not aligned	84
4	2.5 (1)	vents not aligned	22
4	2.5 (1)	vents aligned	18

Room cooling has several advantages:

- There is less pallet rehandling because produce can be cooled and stored in the same place.
- Facility design and operation are simple.
- Peak loads on the refrigeration system are lower than with faster cooling methods.
- Equipment is less expensive than for fast cooling.
- Room cooling works well for unpacked items, which have good exposure to airflow.

Room cooling does have severe disadvantages, however:

- Cooling is slow, often requiring at least 24 hours, or longer, for optimum storage temperature to be reached (see appendix B-1). Palletized boxes with little or no venting may take weeks to cool. Even less perishable perishables, such as apples or pears, may deteriorate unacceptably before they are completely cooled. In some cases, produce respiration may not allow produce to ever reach cold-room temperature. In practice, slow cooling often results in shipping produce before it is adequately cooled, or in a need to delay shipments to achieve proper produce temperature. Sensitive produce can deteriorate measurably in the time required for complete cooling.
- Moisture loss is greater than for faster cooling methods, unless the produce is packed in plastic bags or liners.
- More floor space is needed for a given output than with faster cooling methods.
- Use of the same space for cooling and storage exposes previously cooled produce to high air speeds and fluctuating air temperatures caused by newly added warm produce. Moving cooled produce to a separate storage area involves the same rehandling as for faster cooling methods, but without the advantages of those methods.
- Moisture released by warm produce may condense on colder produce, causing decay growth.
- Cooling rates are variable, even for the same produce item, making produce temperature difficult to manage.

A variation of room cooling is to process and package perishables in a cold environment to minimize the cooling needed after packaging. This is a common practice with fresh-cut fruit and vegetables. Both are commonly packaged in modified-atmosphere bags that prevent cold air in the storage room from making direct contact with the produce after packaging. For example, fresh-cut salad vegetables are sorted and cut in a room kept at about 2° to 4°C (36–39°F). After this step, the vegetables are hydrocooled in an immersion cooler, dewatered in a centrifuge, and packaged in the same refrigerated room. Cherries are hydrocooled and then transported through the packinghouse to the packaging area in flumes filled with refrigerated water. Neither system allows the produce to reach temperatures close to 0°C (32°F), which are needed for the longest produce life (Toivonen 2014). Because of this, some cherry packers in California have recently installed forced-air coolers to finish cooling the packaged fruit.

Forced-air cooling

Forced-air cooling is the most widely used method of cooling fruit, vegetables, and flowers (Guillou 1960; Parsons et al. 1972a, 1972b; Rij et al. 1979; Baird et al. 1988). With this method, cold air is forced through containers and past individual pieces of produce. Cooling times are much shorter than with room cooling. Nearly all fresh-market commodities can be cooled with forced air, but it is commonly used for tree fruit, berries, melons, and cut flowers. Unlike hydrocooling, ice cooling, and water-spray vacuum-cooling methods, forced-air cooling does not require that produce be packed in water-resistant containers. Forced-air cooling has some disadvantages, however: It is usually slower than other methods (except room cooling) and may cause excessive water loss in some commodities.

Configurations for forced-air cooling

There are many configurations for delivering cold air to packed boxes or bins. Each has advantages in terms of capital cost, effective use of floor space, daily capacity, and ease of produce management. A discussion of the four most widely used layouts follows.

Tunnel coolers. The tunnel cooler is the most common design for forcing air through produce in boxes. Pallet loads of produce are placed in two lanes on either side of an open air channel (figs. 1.10 and 1.11). A tarp is placed over the lanes of pallets, covering the open channel to form a tunnel. A fan removes air from one end of the tunnel, pulling cold air through the packed boxes on either side of the tunnel. The warmed air exiting the fan passes through evaporator coils, where it is recooled, before returning to the room. (In the past, ice blocks, rather than evaporator coils, were used to cool the air.) Pallet loads can be stacked two or three high to make better use of the interior volume of the cooler room (fig. 1.12). Produce in pallet bins can also be cooled with this air-management system if bin walls are vented.

This system cools large amounts of produce in a single batch without specifically managing the temperature of individual boxes or pallet loads. Many installations are set up so that each batch of produce has a separate fan. Fans may be fitted with motor-speed control that enables them to spin more slowly as the air exiting the tunnel drops in temperature during the cooling process. Variable fan speed reduces energy use, reduces heat input to the refrigeration system, and may reduce produce moisture loss. Reversing the direction of airflow during the cooling process can speed cooling, especially for small-diameter items like strawberries.

A tunnel-type cooler can be set up by adding a fan and tarping system to an existing cold-storage room (fig. 1.13). The warm air from the fan should be directed upward to the refrigeration coils in the ceiling and kept separate from refrigerated air entering neighboring coolers. The amount of produce cooled will be limited by the refrigeration capacity of the cold-storage room.

Conveyor coolers. High-capacity conveyor coolers have been used extensively for strawberries and other berries in California. They have

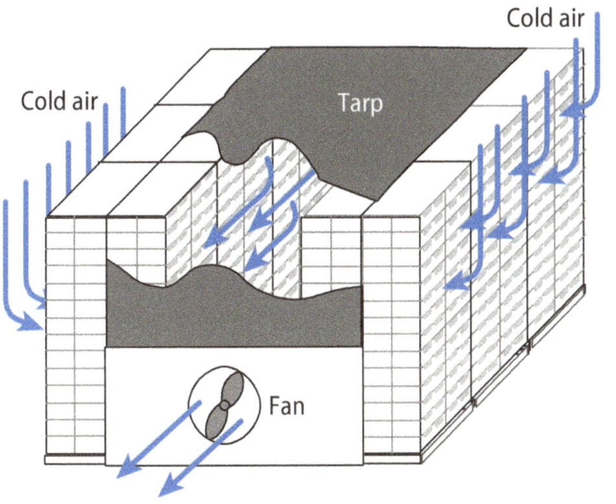

Figure 1.10. Schematic of a tunnel-type forced-air cooler.

Figure 1.11. Tunnel-type forced-air cooler. Pallet openings have been sealed at floor with boards. Deflection of tarp stiffeners and tarp indicate a well-sealed air-return channel. *Photo:* James F. Thompson.

Figure 1.12. Tunnel-type forced-air cooler with pallets stacked two high. *Photo:* James F. Thompson.

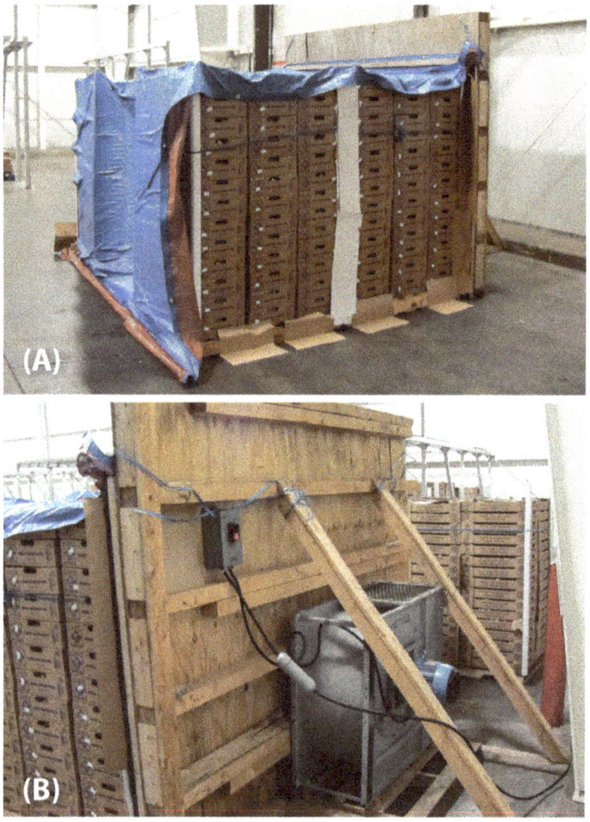

Figure 1.13. Fan unit to implement a tunnel-type cooler in an existing cold room, front view (A) and back view (B). *Photos:* James F. Thompson.

also been used for cauliflower, broccoli, peppers, table grapes, squash, and peaches in bins. Pallet loads are placed on a conveyor that transports them from an outside door, through a cooler divided into several sections, and finally into a cold-storage room (fig. 1.14). At each step in the cooling process, a batch of warm pallets fills the first section of the cooler and the same number of pallets exits the last section of the cooler.

Cooling time is short—about 1 hour for strawberries in single-layer clamshell packages. The fast cooling is caused by very high airflow, approaching 5 liters per second per kilogram (4.8 cfm/lb). In some conveyor coolers, the airflow direction is reversed each time the produce moves to another section as it progresses through the cooler (Hawkins 2010). This design provides more uniform produce temperatures at the end of cooling than do conventional single-direction designs with lower airflow rates. Conveyor coolers also significantly reduce heat load to the refrigeration system by eliminating

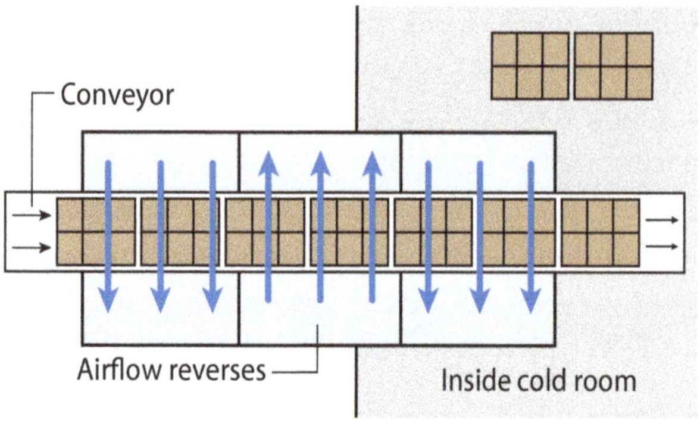

Figure 1.14. Schematic top view of a continuous-flow forced-air cooler.

lift-truck activity, lights, and people in the cooler and by requiring much less refrigerated volume than a traditional tunnel-type cooler. A disadvantage of airflow reversal is that the warmest produce is often in the center of a pallet, where it is difficult to measure its temperature.

Cold-wall coolers. The cold-wall cooler efficiently cools small lots and incomplete pallet loads of produce. Single pallets are placed side by side against a plenum, usually in one row on the floor and in one or two rows above on elevated racks or shelves (figs. 1.15 and 1.16). The interior of the plenum is under suction or, in some coolers, under pressure. Dampers at each pallet position open mechanically with pallet contact, allowing airflow to begin. Dampers can also be operated electrically, controlled with a proximity switch, and can be

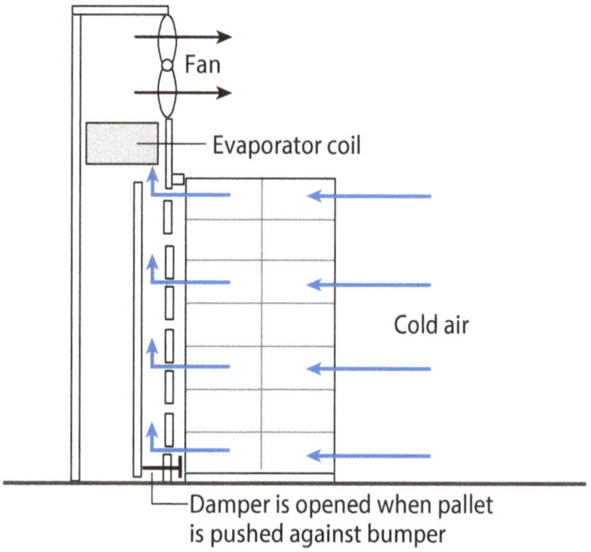

Figure 1.15. Pallet placed against cold-wall cooler.

Figure 1.16. Cold wall–type forced-air cooler.

designed to open only as high as the stack of containers on the pallet, allowing stacks of various heights to be cooled without adjustments or loss of air (fig. 1.17).

Cut flowers are often cooled with pressure-type cold-wall coolers. Packed boxes are placed against a plenum pressurized to about 2 to 4 centimeters (1–1.5 in) of water-column static pressure. The plenum has holes that align with the box vents and are designed to be small enough so that the pressure drop across the holes is relatively large compared with the pressure drop through the box. Airflow through the holes does not change much when the boxes are placed in position, so that the holes are always left open and there is no need for a damper system.

Figure 1.17. Cold-wall cooler for packed cut flowers.

Produce placed in a cold-wall cooler starts to cool immediately. Operators do not need to position tarps or start fans. Every pallet or bin is accessible for removal at any time and may be immediately replaced by another, allowing virtually continuous use of all cooling positions. However, the disadvantage of this system is that each pallet may finish at a different time, forcing operators to monitor produce temperature frequently to prevent unnecessary exposure to airflow after cooling. Some coolers have a timer system to help with monitoring. Damper opening turns on a green light and starts a timer. After a set period, the timer turns off the green light and turns on a red light, alerting the operator to check the temperature of produce in that position to determine whether cooling is complete.

Most early cold-wall coolers were designed with the plenum placed against an interior wall in the cold room, which results in poor use of the floor area (fig. 1.18). One or more plenums can instead be positioned perpendicular to

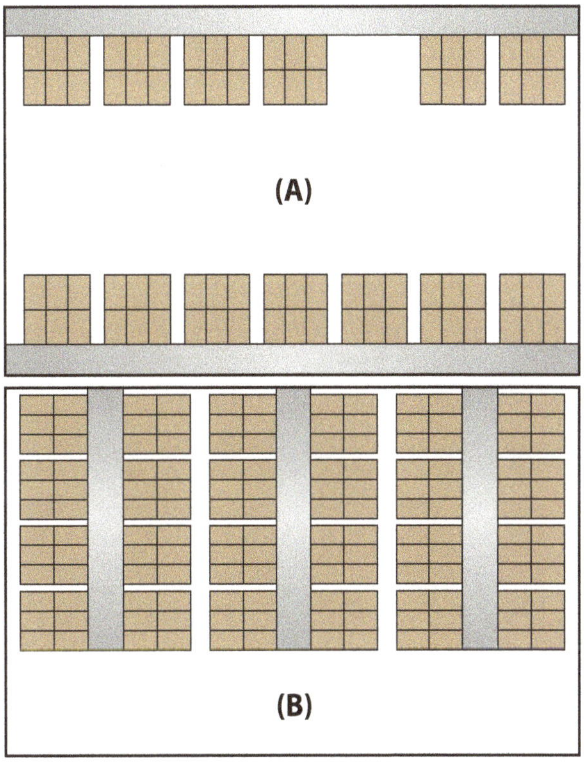

Figure 1.18. Layout of cold-wall forced-air coolers. Air plenums positioned next to room walls (A) and perpendicular to cold-room walls (B).

cold-room walls (see fig. 1.18A; fig. 1.19). This allows both sides of the plenum to be used for cooling and makes efficient use of the cold-room floor area. At relatively little additional cost, plenum fans can be selected that reverse the airflow direction when their rotation is reversed, allowing airflow reversal during cooling. Upper-level pallets are supported on a rack that allows forklift access from the front of the cooler. This type of rack system is also commonly used in forced-air ripening rooms.

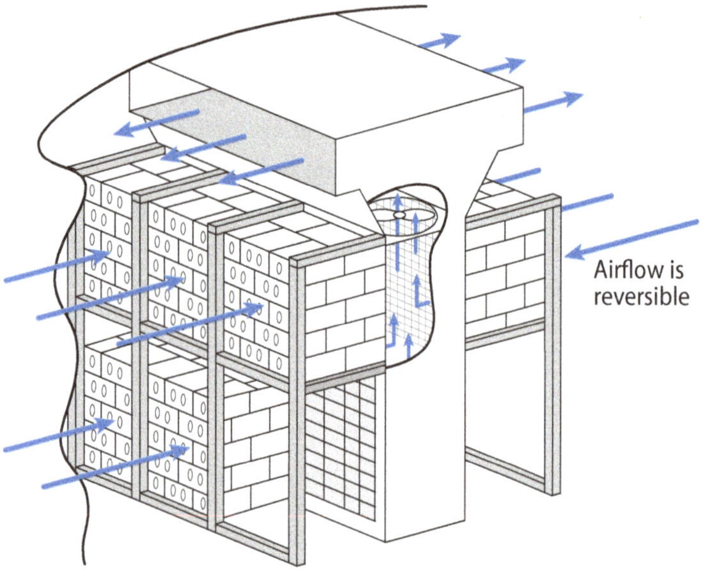

Figure 1.19. Cold-wall forced-air cooler.

Serpentine coolers. The serpentine cooler is designed to cool produce in pallet bins with vented bottoms (fig. 1.20 and 1.21). The openings underneath the bins for the forklift tines are used as air-supply and air-return channels. Stacks of even numbers of bins are placed against a plenum wall, which has horizontal slots that align with the forklift openings on alternate bins. Narrow tarps are placed over the forklift openings that are opposite the plenum wall slits, on the bins farthest from the fan. This arrangement forces cold air to flow in the uncovered forklift openings, through produce in the bins, and back to the plenum wall through the channels formed by the forklift openings.

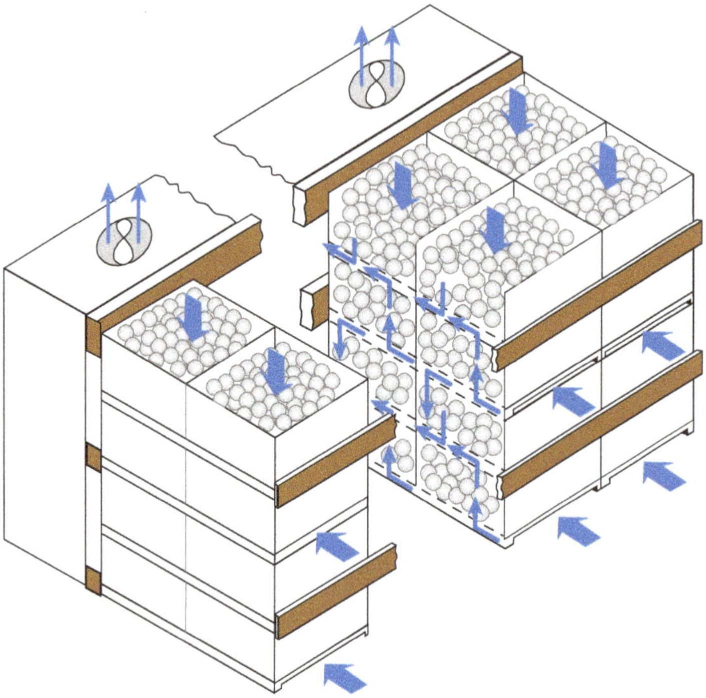

Figure 1.20. Airflow pattern in a serpentine forced-air cooler. This system is specific for produce in bins. By blocking alternate forklift openings on the cold wall, air is forced to pass vertically through bins.

Figure 1.21. Serpentine forced-air cooler.

Because forklift openings allow only a small cross-sectional area for airflow, a serpentine cooler is limited to relatively long cooling times if standard fruit bins, which are 0.6 meters (2 ft) deep, are stacked more than two- or three-deep from the wall. Air speed through bin openings should not exceed 5 meters per second (about 1,000 feet per minute, or fpm), although it may be possible to use air speeds of 7.5 meters per second (1,500 fpm) and still have acceptable variation in cooling times between bins in different positions. Most serpentine systems are used for initial produce cooling prior to packing. Airflow rates are low, often 0.25 to 0.5 liter per second per kilogram (0.24–0.48 cfm/lb) or less. Cooling is often done overnight, and produce is considered adequately cool when it reaches 5° to 10°C (41–50°F). Final cooling is completed after packing.

This system is often operated as a single batch, although airflow can begin as soon as several lanes of bins are stacked against the plenum wall. Serpentine cooling makes good use of cold-room volume.

Cooler layout

Figure 1.22 shows the basic layout of a well-designed cooling and cold-storage facility. Produce from the field arrives at a covered area on one side of the facility. Pallets can be identified, and produce quality can be sampled, in this area. The warm pallet loads are then moved into the cooling area. In tunnel-type coolers, each cooling position should be separated from neighboring positions with a wall. The wall prevents warm produce in a neighboring cooling position, and the warm air it produces, from slowing the final cooling of a batch.

When cooling is completed, produce is moved into a separate area for cold storage. It is difficult to keep the forced-air cooling area at a consistent low air temperature because of frequent door openings and warm produce moving into the area. The last degree or two of cooling can be accomplished with room cooling in the storage area. Stored produce is shipped from a loading dock located next to the storage area and away from the traffic in the receiving area.

Daily produce volume can vary by almost an order of magnitude between the first weeks of the harvest season and the peak of harvest. Dividing the refrigerated volume with insulated walls so that portions of it can remain unrefrigerated when not needed will significantly reduce electricity use during the season.

Cooling time

The average produce temperature during forced-air cooling can be approximated with the pattern shown in figure 1.23. (Produce in a room

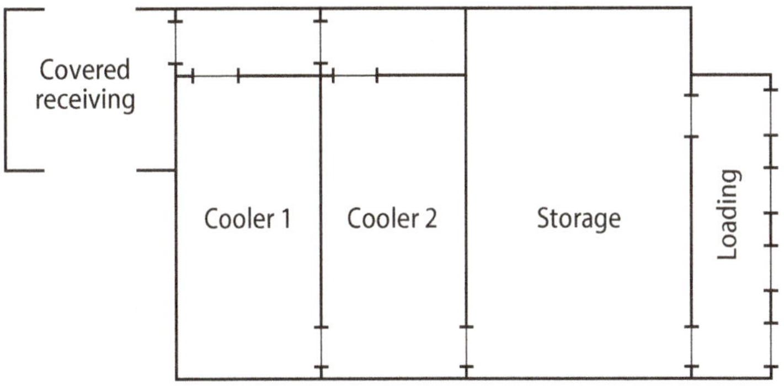

Figure 1.22. Cooler 1 and the storage can be left unrefrigerated when daily cooling need is low and produce is shipped immediately after cooling at the beginning and end of the harvest season.

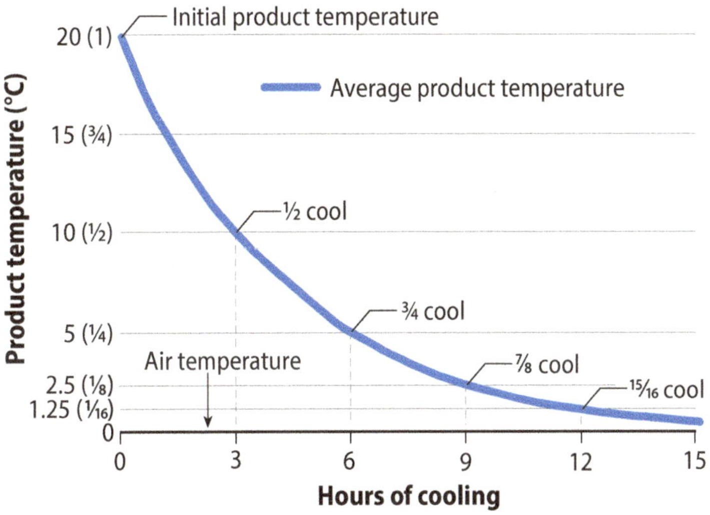

Figure 1.23. Simplified temperature pattern of produce during forced-air cooling. The numbers in parentheses indicate the fraction of temperature drop remaining, obtained by dividing the difference between the produce temperature and cooling-air temperature by the difference between the initial produce temperature and the cooling-air temperature.

cooler or a hydrocooler follow this same temperature pattern.) The rate of temperature drop is related to the temperature difference between the produce and the cold air. Produce temperature drops rapidly at the beginning of cooling; the rate of temperature drop slows as the produce nears final temperature. This process is often approximated with the concept of half-cooling time, the time required for the produce temperature to drop halfway between the initial produce temperature and the temperature of the cold air. In figure 1.23, the produce cools by 10°C, from 20°C to 10°C, in the first half-cooling period. During the next half-cooling period, equal in time to the first, the produce again loses one-half the difference between its temperature at the beginning of the cooling period and the temperature of the cold air. But the temperature drop during the second half-cooling period is half as much as during the first period, only 5°C. Produce is typically left in the cooler for 3 half-cooling periods to achieve seven-eighths cooling, or 4 half-cooling periods to achieve fifteen-sixteenths cooling.

This cooling pattern demonstrates the need to keep cold air close to its set-point temperature near the end of cooling. If the refrigerated air temperature rises only a few degrees during the third or fourth half-cooling periods, produce cooling will slow and may even stop. Tunnel coolers should be built as individual rooms or divided into sections so that warm produce arriving later in the day will not affect the air temperature near batches that are almost cooled. Air temperature can be set at or slightly below the highest freezing temperature of the produce (see appendix A for the highest freezing temperatures of many fruit and vegetables).

Figures 1.24 and 1.25 are more accurate representations of produce cooling patterns. The lengths of the half-cooling periods are not exactly equal, although assuming equal times is a good approximation. The figures also show that produce closest to the cool-air supply takes less time to cool. For large-diameter produce like cantaloupes, the difference between the fastest- and slowest-cooling melons is not great. For small-diameter produce like cherries, however,

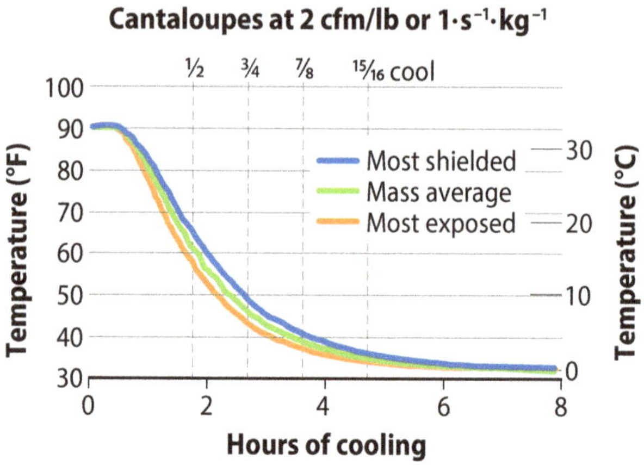

Figure 1.24. Produce temperature in a forced air–cooled pallet of packed cantaloupe boxes.

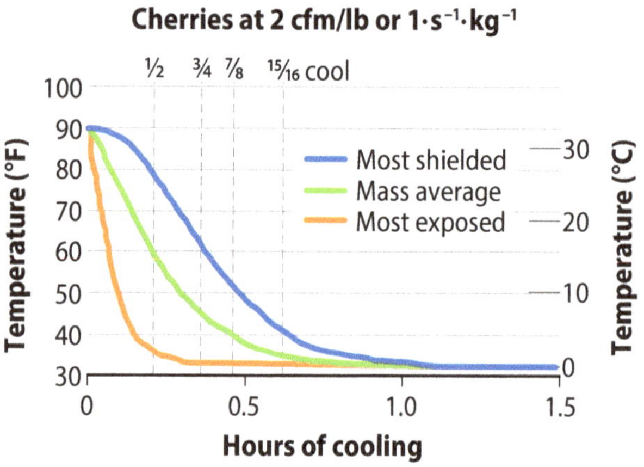

Figure 1.25. Produce temperature in a forced air–cooled pallet of packed cherry boxes.

the difference between the cooling times for the fastest- and slowest-cooling cherries is large, because the relatively large surface-to-volume ratio of cherries allows heat to be removed quickly from the fruit, causing significant warming of the cooling air as it passes through the pallet load. The cooling-system operator must ensure that a load is not removed from the cooler until the warmest produce has reached the desired temperature.

Total cooling time depends on airflow rate and produce diameter. Figure 1.26 shows that a large-diameter fruit takes considerably longer to cool than a small-diameter fruit. A cantaloupe that is 15 centimeters (6 in) in diameter takes 2 to 3 times longer to cool than a table grape with a diameter of only 1.3 centimeters (0.5

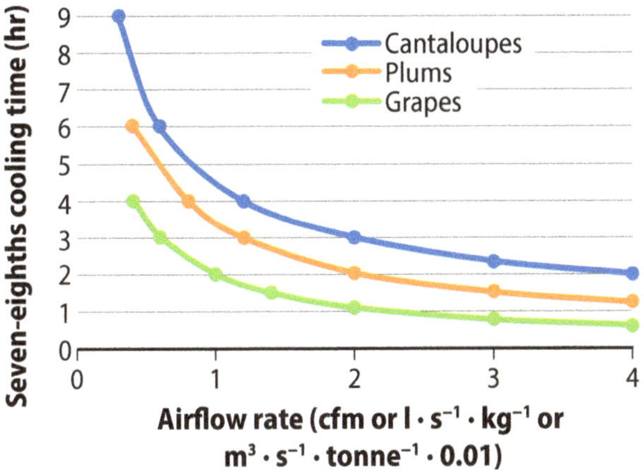

Figure 1.26. Effect of airflow rate on cooling time for cantaloupes in corrugated containers with 6% sidewall venting, plums in corrugated containers with 4% sidewall venting, and grapes in full-telescope corrugated containers with 5.8% sidewall venting.

in). Cut flowers, which usually have very small maximum diameters, have seven-eighths cooling times of 10 minutes to 1 hour at a water-column static pressure of 2.5 centimeters (1 in), producing an airflow of 2 to 4 liters per second per kilogram (1.9–3.8 cfm/lb). Figure 1.26 also shows that increased airflow, measured in air volume per time divided by produce mass, reduces cooling time. Many coolers are designed to provide an airflow of about 1 liter per second per kilogram (0.96 cfm/lb), which results in acceptable cooling times and requires a reasonably sized fan.

However, using higher airflow rates and reduced cooling time allows cooling positions to be used more times during the day and reduces the floor space needed for the cooling operation. Doubling the airflow from 1 to 2 liters per second per kilogram (1.9 cfm/lb) reduces cooling time by about 33 percent for plums and 50 percent for grapes. However, a doubled airflow forces the fan to operate against about four times more static pressure, which requires a motor with about six to seven times more power. The static pressure increase can be minimized by using packages designed with large sidewall vent areas to minimize airflow resistance. Airflow rates of about 4 to 5 liters per second per kilogram (3.9–4.8 cfm/lb) are used in continuous-flow coolers for berries. Strawberries and bushberries are packed in specially designed corrugated fiberboard trays with vent areas exceeding 10 percent.

Fruit and vegetables packed in plastic box liners cool slowly because the liner prevents the cooling air from directly contacting the produce. The optimum airflow rate for this type of packaging is about 0.35 liters per second per kilogram (0.34 cfm/lb) (O'Sullivan et al. 2016).

Airflow rate, pressure drop, and cooling time for a number of selected packaged commodities are listed in appendices C-1 and C-2. The relationship between airflow rate and cooling time is linear when plotted on a log-log graph, which makes it possible to estimate cooling time with limited field data (fig. 1.27).

Width of container stacks for forced-air cooling

Wider produce stacks increase the distance that cooling air must travel through the produce. This usually increases the temperature difference between the coolest and warmest produce at the end of the cooling cycle. If, however, airflow is increased in a wide stack so that airflow per produce mass ($l \cdot s^{-1} \cdot kg^{-1}$ or cfm) remains the same, the temperature difference between the warmest and coolest produce will be about the same as that for a narrow stack. The static pressure required to move air increases rapidly as stack width is increased. Coolers for produce in containers are usually designed for air to flow across the width of one standard Grocery

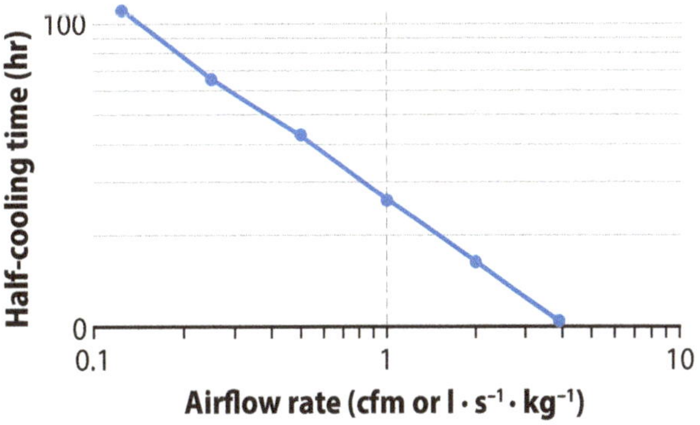

Figure 1.27. Log-log plot of half-cooling time versus airflow rate for a simulated produce item (52mm solid polymer spheres). *Source:* Data from de Castro et al. 2005.

Manufacturers Association pallet, which has a maximum produce width of 1.0 to 1.2 meters (40–48 in).

Fan selection

Fans are selected on the basis of their airflow capacity at a specific static pressure difference between the inlet and outlet of the fan. The static pressure across a pallet is closely related to the sidewall vent area of the boxes in the pallet load, as shown in figure 1.28. Fans producing static pressure differences greater than 5.1 centimeters (2 in) of water column are rarely used because they require high energy use, and in tunnel-type coolers often cause boxes to shift on the pallet during cooling.

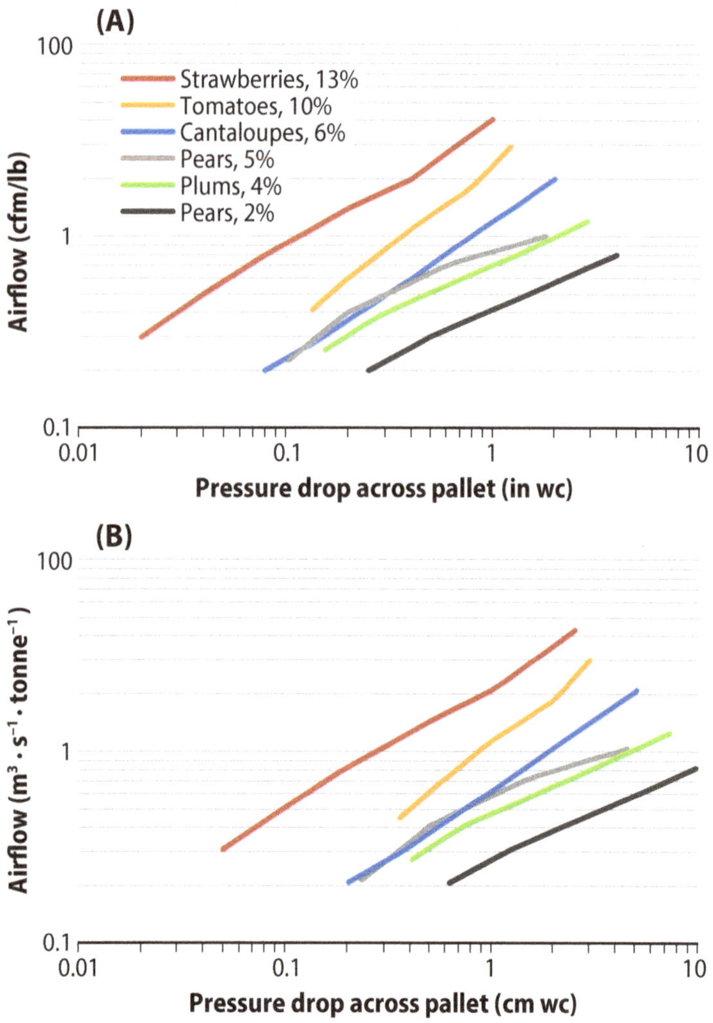

Figure 1.28. Pressure drop across a pallet-width of containers with various levels of sidewall venting in SI (metric) units (A) and customary units (B). Graphs are log-log plots.

Some returnable plastic containers have vent areas in excess of 25 percent of the sidewall area. These large vent areas allow high airflow rates at low static pressures. Their disadvantage is that the static pressure across the pallets can be close to the pressure difference along the air-return channel in a forced-air cooler. This causes fast cooling due to the high airflow through the pallet loads closest to the fan, and slow cooling due to the low airflow through the pallet loads farthest from the fan. In tunnel-type coolers, containers with large vent areas and containers with small vent areas should not be cooled together because of significant differences in cooling time.

Total fan airflow capacity must include air that bypasses boxes and air leaks. Some box designs, such as boxes with chamfered corners or nestable plastic boxes, allow cooling air to flow between boxes, as well as through them. The fan airflow capacity should be increased to account for this. For example, to produce the same cooling times, grapes in wood-end boxes with 2-centimeter (0.75-in) cleats between box layers require twice as much airflow as grapes in boxes without cleat separators. Air can also leak through forklift openings in pallets, gaps between pallet stacks, and poorly placed tarps. Little information is available about air leakage, but it may be as low as 10 percent in a well-operated system and perhaps over 30 percent in a poorly operated one.

A fan for a forced-air cooler should be capable of operating at a range of airflows and pressures because it may be used for a variety of produce items and packaging types. Even if a cooler is specified to operate at a single airflow rate, the actual rate may change when varying amounts of produce are placed on the cooler. Also, as previously mentioned, some coolers are designed to reduce airflow as the cooling process nears the end point. Usually the fans for these coolers are selected to provide the maximum expected airflow at the maximum expected static pressure. To accommodate a range in airflow needs, some coolers are operated with several fans in series or parallel arrangements. Another commonly used method to manage airflow capacity is to operate fan motors with motor speed controllers that produce a wide range of airflow rates.

Total pressure drop across the fan also includes the pressure drop across evaporator coils and in air ducts. As a general rule, forced-air cooling fans should be selected to provide airflow against a maximum of 5 centimeters (2 in) of water column, unless specific testing indicates lower pressures are acceptable. Axial-flow (propeller) and centrifugal (squirrel-cage) fans are both used for forced-air coolers. Most axial fans are more efficient for systems in which the fan operates against less than 5.1 centimeters (2 in) of water column. Plug fans, which are centrifugal fans without a housing, are commonly used in forced-air coolers because they are quieter than axial fans and distribute air across evaporator coils placed above them.

Fan horsepower requirements are related to airflow rate and static pressure across the fan with the following approximate formulas:

In SI (metric) units,

$$P_f = (q \cdot p)/(\varepsilon \cdot 1{,}000)$$

where

P_f = fan power (kW)

q = airflow ($m^3 \cdot sec^{-1}$)

p = pressure (Pa)

ε = fan efficiency (properly selected fans have efficiencies in the range of 0.4 –0.7).

In customary units,

$$P_f = (q \cdot H)/(\varepsilon \cdot 6365)$$

where

P_f = fan power (horsepower)

q = airflow (cfm)

H = total pressure (in inches of water column and often approximated as static pressure across fan)

ε = fan efficiency (properly selected fans have efficiencies in the range of 0.4–0.7).

Containers for forced-air cooling

The packing system and the containers must permit a satisfactory volume of airflow with a reasonable pressure difference across the stack (Mitchell et al. 1971; Wang and Tunpun 1968; de Castro et al. 2005; Ferrua and Singh 2009).

The area of container vent holes in the sides through which air flows should equal about 5 percent of the side area. Smaller vent areas restrict airflow, causing longer cooling times. Larger vent areas can speed cooling, but large vent areas also allow faster produce warming if the box is removed from the cold chain. Large vent areas also tend to weaken corrugated containers. Most corrugated fiberboard containers can have up to 5 percent venting without being specially designed for additional strength. Trays commonly used for berries and some vegetables have been designed with vent areas exceeding 13 percent of the side wall area.

Box liners and consumer packages, like plastic clamshell boxes and bags, restrict cooling air from flowing directly past the packed produce and cause significantly longer cooling times than for produce packed without interior packaging (Ngcobo et al. 2012). Liners should be folded over and tucked into the sides of the box to allow cooling air to flow over the top of the liner in each container. Containers should be vented with edge vents to allow air to enter and exit at the top of the container. Microperforations in box liners speed forced-air cooling, although the perforations also allow increased moisture loss of the perishable in subsequent handling. The use and design of interior packaging is based on many postharvest handling constraints other than speed of cooling. For example, 1-pound clamshell packages for strawberries are sometimes designed with 5-percent vent area, even though forced-air cooling is faster with vent areas between 7 and 13 percent.

Fruit packed in paper or plastic trays will cool satisfactorily if the containers are designed to let air pass both over and under each tray. Grapes packed in vented, lidded containers cool well with forced air, despite their compact mass (Luvisi et al. 1995).

Box venting should be designed according to the following general rules:

- Vent size and shape should not allow vents to be blocked by produce.
- Use edge vents on horizontal edges of corrugated boxes if interior packing materials, like box liners or consumer bags, block vents located near the center of box panels.
- Use a few large vents instead of many small vents. Vents should be 1 centimeter (0.5 in) wide or greater.
- Keep vents 4 to 7 centimeters (1.5–3 in) away from vertical edges to minimize strength loss in corrugated containers.
- Vent area should be about 5 percent of the sidewall area. Smaller venting can be used if cooling is slowed by liners or other interior packaging.
- If boxes are to be cross-stacked, use the vent arrangement illustrated in figures 1.29 and 1.30.

Moisture loss in forced-air cooling

Forced-air cooling causes virtually no moisture loss under some conditions. In other cases, losses are great enough to prevent the use of this cooling method. The amount of moisture loss depends on the initial temperature of the produce, its transpiration coefficient, the use of waxes or moisture-resistant packaging, the humidity of the cooling air, and exposure to unnecessary airflow after cooling is complete.

High initial produce temperature causes more weight loss than low initial temperature (fig. 1.31). Initial produce temperature can be kept low by harvesting early in the morning or at night, shading produce that must be temporarily stored in the field, and starting the cooling process as soon as possible after harvest. Some produce items are stored temporarily and partially

Figure 1.30. A well-vented box for pears.

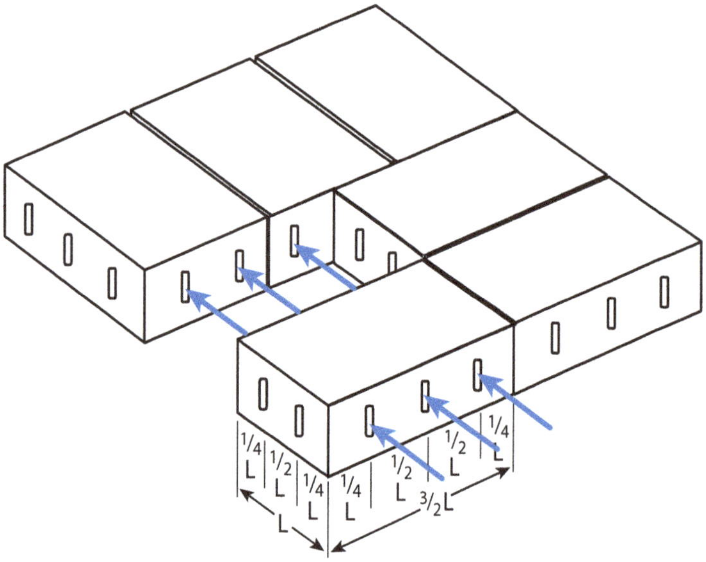

Figure 1.29. Air vent design allowing good airflow with cross-stacked boxes.

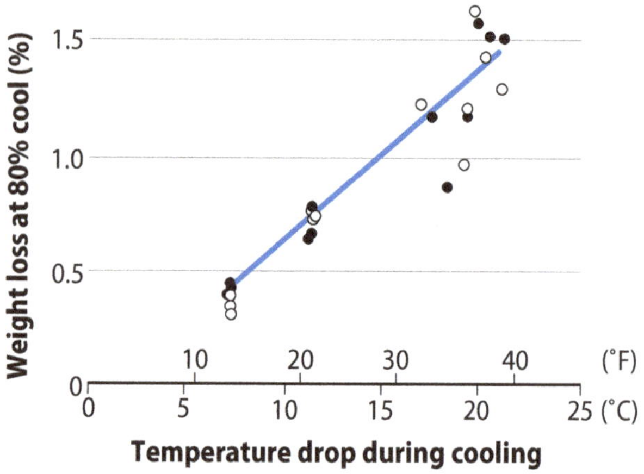

Figure 1.31. Effect of carrot temperature reduction on weight loss during forced-air cooling. Open points are cooling with an airflow of 0.5 cubic feet per minute per pound and closed points are cooling with 2.0 cubic feet per minute per pound.

cooled before packing, and then cooling is finished after packing. Minimize reheating during packing to reduce moisture loss.

Produce items like oranges, which have a low transpiration coefficient (see table 1.4), lose very little water during cooling and are well suited to forced-air cooling. The transpiration coefficient of carrots is 10 times greater than that of oranges, and carrots can lose 0.6 to 1.8 percent of their uncooled weight in the tests described in figure 1.31. Packing carrots in polyethylene bags reduces moisture loss to 0.08 percent, although cooling times are up to five times longer than for unbagged carrots. Apples, pears, kiwifruit, and grapes are sometimes packed in plastic-film box liners to retard moisture loss. The liners are folded over at the top and are not tightly sealed. They sometimes have a few holes to reduce moisture condensation. Produce items suited for liner packing have slow respiration rates and can withstand slow cooling. Liner-packed grapes must also have a sulfur dioxide–generating pad inside the liner to prevent decay growth (Gentry and Nelson 1964).

The data in figure 1.31 show that total moisture loss is not affected by the greater airflow rates used for faster cooling. Flockens and Meffert (1972) and Gan and Woods (1989) drew the same conclusion from heat-transfer modeling and testing on other fruit and vegetables. Higher air speeds increase the transpiration coefficient, but the increase is balanced by shorter cooling times.

If the relative humidity of the cooling air is above 80 percent, the effect on moisture loss is negligible (Gan and Woods 1989). Most of the vapor pressure difference between the cooling air and the produce is a result of the warm temperature of the produce, not the humidity of the cooling air. An advantage of high humidity during cooling is that it may add moisture to fiberboard containers and reduce produce moisture loss in subsequent refrigerated handling, because the packaging material will not absorb as much moisture from the produce. However, corrugated fiberboard loses strength as it gains moisture, so the box design will need to provide adequate strength at elevated moisture contents.

A forced-air cooler should be designed and operated to reduce or stop airflow through the produce as soon as it is cooled. Continued flow may cause noticeable water loss from produce unless the air is nearly saturated. Some cooler fans are controlled on the basis of return-air plenum temperature. In this type of system, fans slow when the plenum-air temperature is very close to the supply-air temperature. Less air is needed near the end of cooling than at the beginning because the produce's heat-loss rate decreases greatly near the end of cooling.

Air-channel dimensions in forced-air coolers

Narrow air-supply and air-return channels in tunnel coolers cause uneven cooling (Guillou 1963; Haerter 1963). The channels should be wide enough that air speeds are less than 7.5 meters per second (1,500 fpm). A narrow air-supply channel (fig. 1.32) will cause produce in the bottom boxes in pallet loads to cool more slowly than the top boxes, and a narrow air-return channel will cause the pallets farthest from

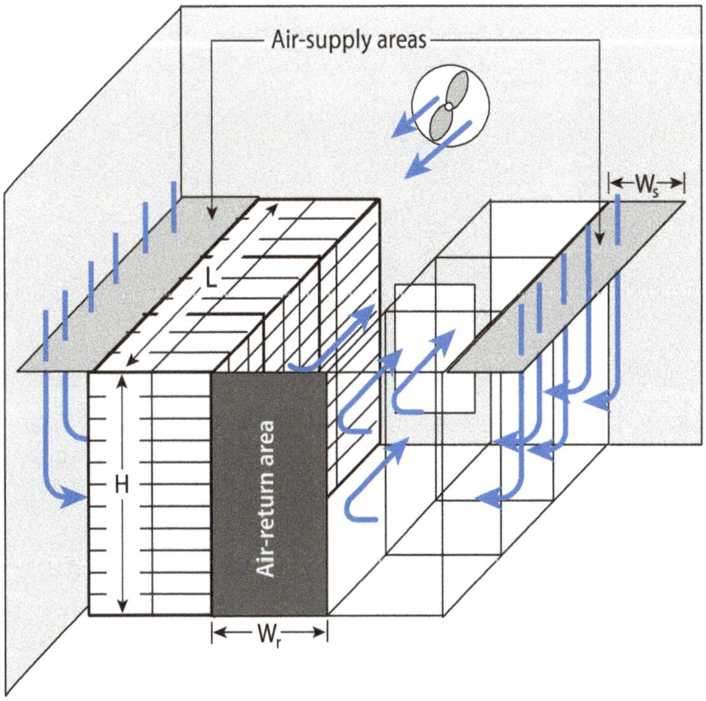

Figure 1.32. Design of air-supply and air-return plenums in a tunnel-type cooler.

the fan to cool more slowly than the pallets closest to the fan.

In addition to measuring temperature variation, another way to identify poor air-channel design is to measure the pressure drop across pallets at several heights in several pallets—and at the same height for pallets at various distances from the cooling fan. Large differences in pressure drop indicate significant differences in airflow through pallets. Lower airflow causes longer cooling times.

The following equations can be used to calculate optimal air-channel widths. The distance between pallet loads in neighboring cooling positions is two times W_s.

$$W_s = Q/(L{\cdot}2{\cdot}S)$$

$$W_r = Q/(H{\cdot}S)$$

where

Q = fan air volume (m^3·sec^{-1} or cfm)

L = length of pallets on cooler (m or ft)

W_s = air-supply channel width (m or ft)

H = height of pallets on cooler (m or ft)

W_r = air-return channel width (m or ft)

S = maximum air speed (7.5 m·sec^{-1} or 1,500 fpm).

Ideal air-channel widths can also be determined using figure 1.33, which is based on the above equations.

Reducing electricity costs

An energy accounting of forced-air coolers (Thompson et al. 2010) showed that a seasonal average of only 36 percent of the electricity consumption was for produce cooling (table 1.7). The rest was used for removing heat from fans, lights, heat conducted across exterior cooler surfaces, and lift trucks.

Heat input based on seasonal average data hides the effect of produce throughput on efficiency. During a 7-month season, a strawberry cooler had fruit heat input that ranged from 35 percent of total heat input, for the month with the lowest produce throughput, to 70 percent for the month with the highest throughput (Thompson et al. 2010). Refrigeration required to keep the cooler building at set-point air temperature was relatively constant during the

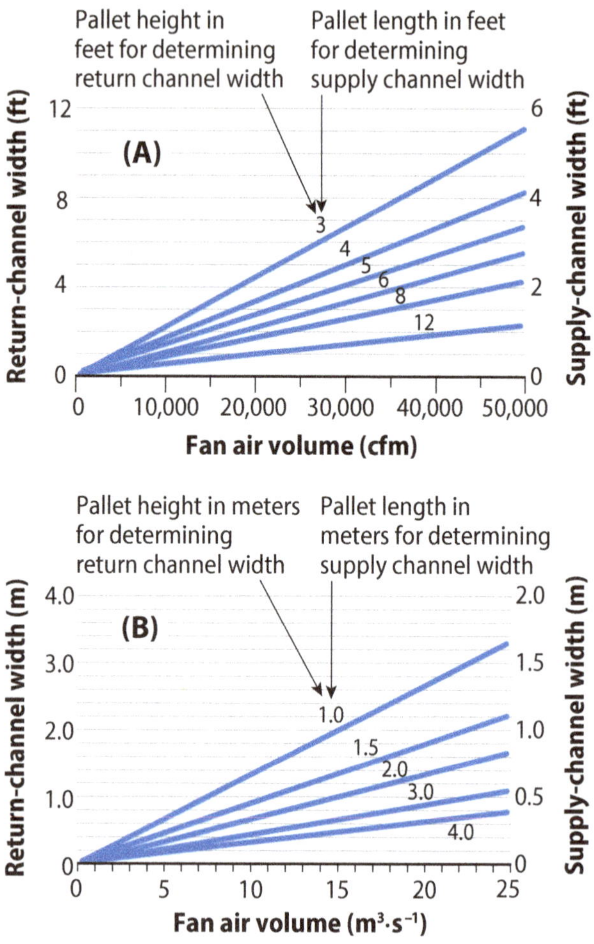

Figure 1.33. Recommended dimensions of air-supply and air-return plenums in a tunnel-type forced-air cooler, in (A) customary units and (B) SI units.

Table 1.7. Season average electricity use for major heat inputs to forced-air coolers*

Heat source	Average electricity use (% of total)
product	36
fans	28
lights	17
building walls and roof	14
electric lift trucks	6

*Includes electricity for operating equipment and for refrigeration needed to remove heat from the product or the equipment.

cooling season and was the majority of the heat input during the months with low produce throughput. This resulted in monthly energy coefficients ranging from 0.30 to 1.25 (fig. 1.34). High energy efficiency can be maintained during the entire season by dividing the cooling capacity into sections and shutting down sections that are not needed. Conveyor-type forced-air coolers require a very small building and their refrigeration can be easily turned off when not needed. There is no published electricity-use data on these coolers, but they likely operate with high energy efficiency.

Additional options for reducing electricity use in forced-air coolers include

- cooling after culling to avoid cooling fruit that will not be sold
- reducing fan energy use by using well-vented boxes; selecting an evaporator coil with a low pressure drop across it; turning off or slowing fans when not needed for cooling; selecting fans with good efficiency under typical air pressures and flow rates; and using high-efficiency motors
- using high-efficiency light sources, like high-bay fluorescent lamps or LEDs
- insulating the cooler (new coolers are often built with R-40 insulation in the walls and R-60 in the ceiling); using high-reflectivity roofing and exterior paint
- immediately removing produce from cooler when it reaches desired temperature
- restricting air infiltration by using fast-acting doors, flap doors, and seals between the cooler building and refrigerated transport vehicles
- efficiently sequencing screw compressors
- minimizing head pressure with high–surface area condensers equipped with fan speed control
- maximizing suction pressure with a floating pressure-control system

Table 1.8 compares the energy efficiency of forced-air cooling with four other cooling methods, based on a number called an energy coefficient. The coefficient equals the cooling work done divided by the energy purchased to operate the cooler (Thompson and Chen 1988). High values represent high efficiency. Vacuum cooling is the most energy-efficient, followed by hydrocooling, water-spray vacuum cooling, package icing, and forced-air cooling. Part of the reason for the high efficiency of vacuum cooling is that it removes heat only from the produce being cooled. The other types of coolers remove heat from fans, pumps, infiltration of outside air, heat conducted through exterior walls, lights, forklifts, and people working in the cooler.

The energy coefficients of coolers, even of the same type, can vary considerably. For example,

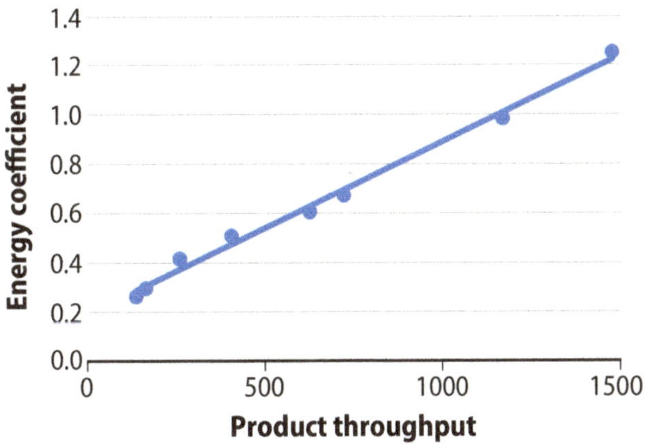

Figure 1.34. The effect of monthly produce throughput on monthly energy coefficient for a strawberry forced-air cooler.

Table 1.8. Comparison of season average energy efficiency of commercially operated coolers handling commodities typical for the cooler type, listed in order of increasing average energy coefficient

Cooling method	Average energy coefficient	Range
forced air	0.4	0.1–0.7
package icing	0.7	0.3–1.0
water-spray vacuum	1.1	0.9–1.3
hydro	1.4	0.7–2.3
vacuum	1.8	1.5–2.5

Source: Adapted from Thompson and Chen 1988; Thompson et al. 2010.

a well-operated hydrocooler can operate more efficiently than most vacuum coolers. However, a poorly operated hydrocooler is not much more energy-efficient than many forced-air coolers.

Use of expensive peak-time electricity can be reduced by installing an engine generator set on-site; slowing produce cooling to shift refrigeration demand to later in the day; using electric lift trucks and charging their batteries at night; and installing ice-building equipment, which allows refrigeration equipment to operate at night and store ice for use the following day (Thompson and Knutson 1989). The financial feasibility of generating power on-site depends mainly on the local cost of electricity and fuel. Facilities in areas where electricity costs are high can pay off the investment in equipment after a few seasons of electricity savings. Air pollution regulations restrict the use of stationary engines in some regions. Using less airflow and slowing cooling can work well if a facility has unused cooling capacity at night. The quality of most commodities is not adversely affected if produce cooling begins quickly after harvest and final cooling is postponed to night hours. Battery-operated lift trucks reduce electricity demand because they produce less heat than propane-powered units, and batteries can be charged during off-peak hours. Fast charging and opportunity charging have added more options, but the goal is still to minimize charging during the peak-rate period of the day. Even though the initial investment for electric lift trucks is typically higher than for propane-powered models, the lifetime cost of ownership is 30 percent less for electric trucks with lead-acid batteries and 75 percent less for trucks with lithium-ion batteries, because of lower maintenance and energy costs. In addition, future technical improvements may further decrease the energy consumption of electric trucks. Propane lifts should not be operated inside enclosed spaces because they expose workers to carbon monoxide and release ethylene, which can damage some perishables. Packed-tower refrigeration systems, which use cold water to cool air, are well suited to ice-building equipment, but they are not easily set up to cool air below 0°C (32°F), as is often needed in forced-air coolers. Cold-wall systems can be set up to force the air directly from a packed-tower heat exchanger to the produce, preventing the cooling air from warming before it reaches the produce, as it does in tunnel coolers.

Troubleshooting slow-cooling problems

Slow cooling in tunnel-type forced-air coolers is caused by inadequate refrigeration capacity, poor airflow management in the refrigerated space, or inadequate airflow through boxes.

Symptom 1: Cooling times increase excessively during periods when large amounts of produce are being cooled.

Cause: Refrigeration capacity is inadequate.

Undersized refrigeration equipment or poor equipment maintenance may allow the temperature of the cooling air to rise. Work with a refrigeration engineer to evaluate equipment performance and determine the best, least expensive way to increase refrigeration capacity.

Symptom 2: Temperature of the cooling air rises above the set-point temperature near the end of a cooling cycle.

Cause: Recently added warm produce is heating the cooling air in a neighboring position that is nearing the end of a cooling cycle.

In tunnel-type coolers, warm produce entering the cooler, or warm air infiltrating into the refrigerated volume, may cause a temporary increase in air temperature in some parts of the cooler, thereby slowing cooling or even warming produce that is nearing completion. Use fast-acting doors, strip curtains, or air curtains to restrict air infiltration and do not locate cooling positions next to exterior doors. Divide the cooler into separate air spaces (fig. 1.35) to prevent the air in one cooling position from affecting a neighboring position. If each cooling position in a large room has its own evaporator coil, each position can be set up as a separate airspace. If several positions are cooled with the same evaporator, it may be best to have one airspace per evaporator. Uninsulated walls or even curtains can be used to divide the cooler.

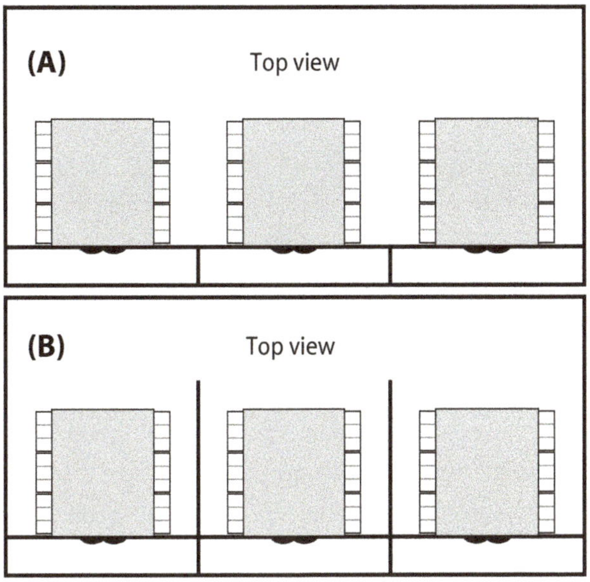

Figure 1.35. In a tunnel-type cooler without dividers (A), warm air from one cooling position can mix with air from other positions. Uninsulated walls can be used to separate cooling positions (B) and prevent warm air from a recently loaded position from slowing the cooling of a neighboring position.

Symptom 3: Temperature records show that cooling-air temperature is at the set-point air temperature during the last two-thirds of each cooling cycle, but some or all boxes cool slowly.

Cause: Not enough air is flowing through boxes. Some possible reasons are

- boxes are not adequately vented (5 percent carton-wall venting is needed, and vents must align when boxes are palletized)
- packing materials are blocking airflow
- fan has too little airflow capacity (1–2 $l \cdot sec^{-1} \cdot kg^{-1}$ [1–2 cfm/lb] is typically needed)
- too many boxes have been placed in the cooler
- air is bypassing boxes through lift truck openings or between poorly positioned pallets, or the tarp on a tunnel-type cooler is not positioned for a good air seal
- boxes with large vent areas are placed in the same cooler as boxes with small vent areas, causing little air to flow through the boxes with less vent area

Fan capacity, measured in cubic meters per second ($m^3 \cdot sec^{-1}$) or cubic feet per minute, can be estimated by measuring the static pressure difference between the fan inlet and outlet and the fan's rotational speed, and then using the fan manufacturer's performance data. A permanently installed pressure gauge measuring return-air plenum pressure will alert operators to airflow problems. For example, air bypassing boxes is associated with unusually low air pressure in the return-air duct.

Hydrocooling

Hydrocooling is used for many types of tree fruit, melons, root vegetables, and stem vegetables (Bennett 1963, 1960; Stewart and Lipton 1960; Mitchell et al. 1972; Pentzer et al. 1936; Perry and Perkins 1968; Toussaint et al. 1955). Hydrocooling is not used for grapes, cut flowers, and most berries because the surface water that remains after cooling encourages decay. Apricots can show damage from chlorine in the hydrocooling water. Most leafy vegetables can be hydrocooled, but they are usually vacuum-cooled or water-spray vacuum-cooled because these methods are faster and vacuum cooling does not require a water-resistant box. Hydrocooling removes no water from the produce and may even revive slightly wilted produce. Effective hydrocooling requires that cooling water

- flow over the surface of the produce
- contact as much of the produce surface as possible
- be kept just below the desired final temperature of the produce
- be kept free of decay-causing organisms

Configurations for hydrocoolers

The two main types of hydrocoolers are water-shower and immersion designs. In the first type, water is pumped above the produce and rains down past it. Produce, whether packed in containers or unpacked, moves slowly and continuously through the shower (figs. 1.36 and 1.37); alternately, produce in containers can be loaded as a batch in the cooler (fig. 1.38).

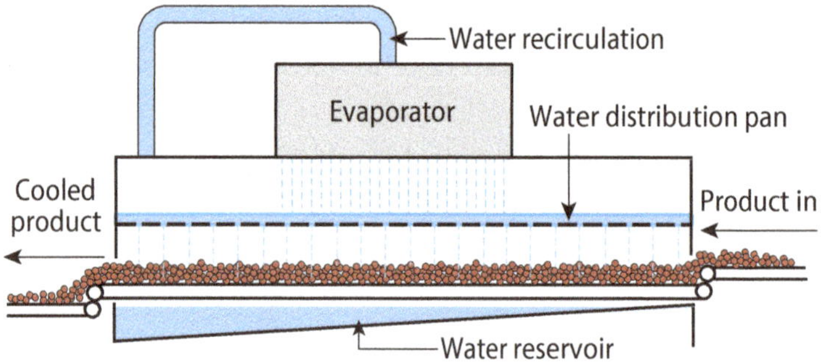

Figure 1.36. Cutaway side view of a continuous-flow shower hydrocooler.

Figure 1.37. In-line shower hydrocooler for cherries.

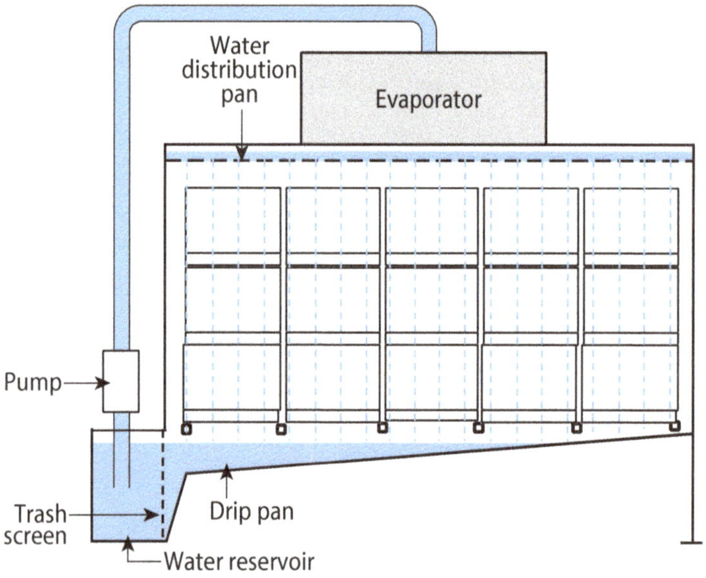

Figure 1.38. Cutaway front view of a batch hydrocooler with produce in bins.

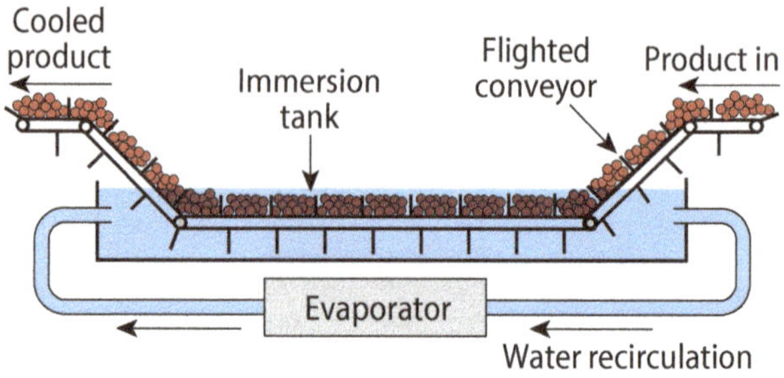

Figure 1.39. Cutaway side view of a continuous-flow immersion hydrocooler.

Shower-type coolers have a drilled metal pan that distributes cold water over the top of the produce. Spray nozzles are sometimes used to distribute water, but they require much more water-pump energy than does a shower-pan distribution system.

Immersion coolers (fig. 1.39) are used mainly for bulk produce. They are best suited to produce, like cherries, that are denser than water and stay completely immersed. They are commonly used in fresh-cut vegetable operations. A conveyor with raised cleats or flights is used to move produce through the cold-water bath.

Cooling time in shower-type coolers fits the logarithmic cooling pattern seen in forced-air cooling but is faster than forced-air cooling (Fricke 2006). Cooling time depends mainly on produce diameter (fig. 1.40). Small-diameter items such as asparagus and radishes have

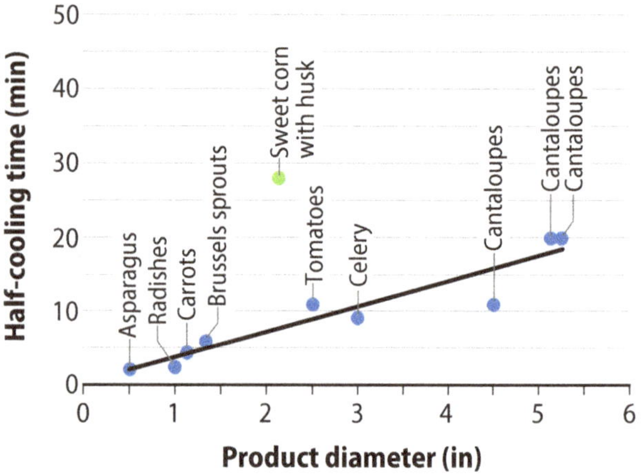

Figure 1.40. Effect of minimum produce diameter on mass average half-cooling time in shower-type hydrocoolers with water flow greater than 10–14 l · sec^{-1} · m^{-2} (15–20 gpm/ft^2). *Source:* Data is derived from Stewart and Couey 1963 and Stewart and Lipton 1960.

half-cooling times of 2 to 5 minutes, whereas a large cantaloupe has a half-cooling time of 22 minutes. A leafy covering, like the husk on sweet corn, causes longer cooling times than would be expected from a solid item with the same diameter. The leaf layers in cabbage also retard heat transfer, causing a head with a diameter of 13 centimeters (5.5 in) to have an 81-minute half-cooling time (Stewart and Couey 1963).

Cooling time for immersion hydrocoolers also depends on produce diameter (fig. 1.41). Based on figures 1.40 and 1.41, immersion cooling time for produce greater than 8 centimeters (3 in) in diameter is about twice as long as with shower-type hydrocoolers. There is little difference in cooling time for fruit and vegetables less than 4 centimeters (1.5 in) in diameter.

Cooling time under commercial conditions will be 4 to 5 times the half-cooling data (15/16ths to 31/32nds cool) shown in figures 1.40 and 1.41. Hydrocooler water is rarely cooled much below 0.5°C (33°F)—and items that need to be cooled close to this temperature will need long exposure times. In fact, it is rare for produce to be hydrocooled below about 1°C (34°F). Some California cherry-packing operations use a two-stage cooling process to ensure their fruit is cooled below 0.5°C (33°F). The fruit is hydrocooled before packaging to about 1 to 3°C (34–38°F) and then master containers of packed cherries are cooled by forced air to below 0.5°C (33°F). Two-stage cooling can also be useful for ensuring low temperature of fresh-cut produce. Two-stage cooling can often be conducted in an existing cold-storage room because the produce is nearly at final temperature and produces little additional heat load.

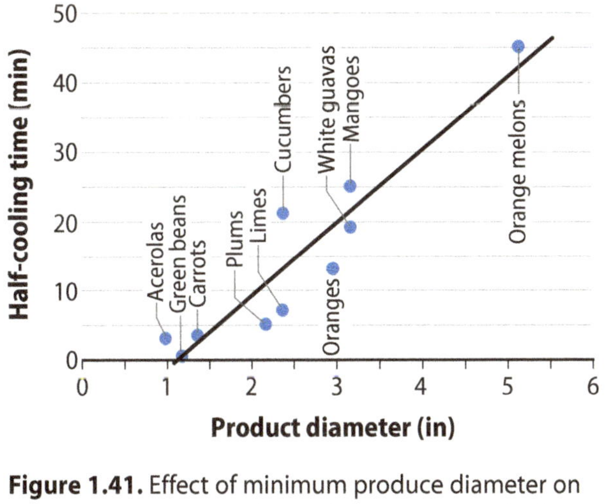

Figure 1.41. Effect of minimum produce diameter on mass average cooling time in immersion-type hydrocoolers, with a water approach speed of about 0.033 m $\cdot s^{-1}$ (6.5 fpm). *Source:* Derived from Teruel et al. 2001.

In shower-type coolers, water flow for produce in shallow layers needs to be in the range of 4.8 to 6.8 $l \cdot sec^{-1} \cdot m^{-2}$ (7–10 gal per minute per square foot of cooling area [gpm/ft^2]) to achieve fastest cooling. As produce depth increases, particularly when produce is in boxes or bins, water tends to flow around certain locations, causing slow cooling in these areas. Increasing the water flow ensures that the locations with restricted flow get enough water flow. Commercial coolers designed to handle double-stacked pallet bins (with a total produce depth of about 1.2 m [4 ft]) use rates of 13.6 to 17.0 $l \cdot sec^{-1} \cdot m^{-2}$ (20–25 gpm/ft^2). Water-flow rates above this level rarely reduce cooling time, and there is little temperature difference between the top and bottom layers of produce at these rates. Batch coolers need the same water-flow rates as continuous-flow shower coolers.

Uniform rapid cooling can be obtained at flow rates lower than those recommended above if container venting is limited so that water nearly fills the container during cooling. Because of difficulties in adjusting and maintaining bottom vent openings to allow water filling, it is easier to speed cooling by increasing the water-flow rate. Wetting agents added to the cooler water have been shown to have little effect on speeding cooling rates (Stewart and Lipton 1960; Bennett 1963).

Hydrocoolers should be designed so that the distance between the water distribution pan and the top fruit or vegetables never exceeds 15 to 20 centimeters (6–8 in). Leafy vegetables, as well as vegetables such as broccoli and asparagus, are subject to water-beating damage if the water-drop height exceeds this range. The damage appears as dark water-soaked areas on the exposed leaves and broken florets on broccoli

Figure 1.42. Water-beating damage to broccoli. *Photo:* Evett Kilmartin.

(fig. 1.42). A water-drop height of more than 20 centimeters (8 in) can cause surface pitting of sweet cherries. Install the water-distribution pan as close to the produce as possible. If the produce height in a cooler is not consistent, a second, adjustable-height pan can be installed below the fixed pan. Produce in pallet bins or field boxes can be covered with a perforated lid to reduce effective water-drop height.

Rapid cooling in immersion coolers requires a water approach speed of about 0.076 $m \cdot s^{-1}$ (15 fpm). Because motion of the produce through a tank is seldom sufficient to achieve this speed, propellers or pumps must be installed to circulate the water. Some coolers use an air-bubbling system to agitate the water. This technique is energy-inefficient unless the air, which cools as it rises through the tank, is recirculated or used to cool a storage area around the cooler.

Water temperature should be kept at 0° to 0.5°C (32–33°F) for most items. Even chilling-sensitive produce that is cooled only to temperatures above 5°C (41°F) can be cooled with 0°C (32°F) water if cooling time is limited. These chilling-sensitive produce types are not usually injured if their average flesh temperature does not drop below their damage threshold (Gaffney and Baird 1975).

Containers for hydrocooling

Containers must have adequate top and bottom venting to allow cooling water to flow through the package. This is a particular problem in hydrocooling items packed in corrugated fiberboard containers, especially when containers are palletized. Vent holes must align even when the boxes are cross-stacked. Leaves on vegetables such as beets or carrots with tops can cause water channeling or can block poorly designed vents, thereby slowing cooling.

Containers and packing materials must be water-resistant. Wood and plastic work well in hydrocoolers. Corrugated fiberboard can be used if it has been hot wax–dipped. Even hot wax–dipped fiberboard, however, sometimes fails if it is exposed to long hydrocooling times. Waxing increases container cost, sometimes to nearly half the total cooling charge, and inhibits the recyclability of the boxes.

Sanitation

Hydrocooler water must be relatively free of microorganisms that cause plant decay and human disease (Ritenour and Crisosto 1996). Water supply should be potable, from a clean well or domestic supply. Water from streams or ponds is rarely clean enough. Because nearly all coolers recirculate water, the cooler must be designed to control disease organisms that enter the system from incoming produce. Extremely dirty produce should be washed before entering the cooler. In some cases—when the water is not recirculated and the water supply is cool—washing may also be used for preliminary cooling.

Hydrocooler water needs to be continuously treated and monitored to minimize the spread of human and plant pathogens. Microbial pathogens arrive on produce and can be redistributed to noninfected items through the cooling water. Pathogens should be controlled with locally permitted chemicals or nonchemical alternatives. The disinfection processes work best when suspended particulates are minimized through filtration. Water must be regularly monitored to ensure that adequate levels of disinfection chemicals are present and that the concentration of chemicals and suspended solids that reduce the effectiveness of the disinfectant are lower than acceptable thresholds.

The cooler should be designed for easy and thorough cleaning. All parts of the water-distribution pan should be accessible. If the pan is covered with insulated panels, the panels should be easily removable to permit convenient inspection for plugged holes. The reservoir must also be accessible so that sediment and debris can be removed when the water is changed. The drain should be large enough for fast draining and placed low so that all of the water drains out. Self-cleaning trash screens placed just ahead of the water pump are used in most hydrocoolers.

Chlorine is often used to disinfect hydrocooler water and is corrosive. Hydrocoolers are often made of stainless steel and plastic parts, which resist chlorine attack. If mild steel is used, it must be painted to provide protection, although paint requires regular maintenance.

Conveyor and water flow design

The hydrocooler conveyor capacity can be estimated from produce cooling time and the rate of produce throughput. Use figure 1.43 or the following formula:

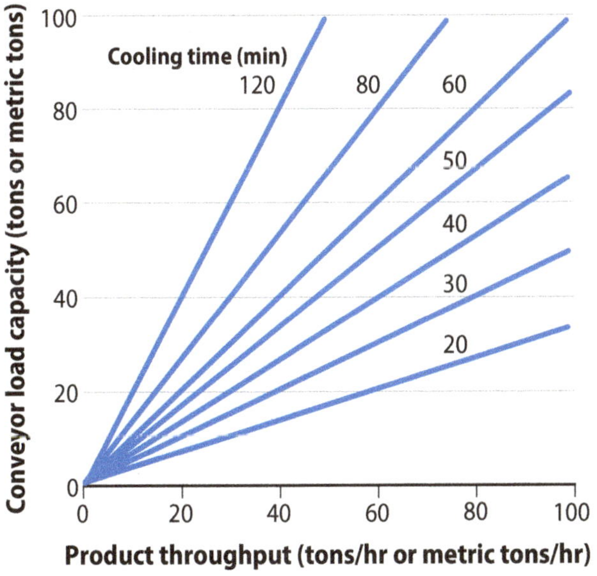

Figure 1.43. Conveyor load capacity for a continuous-flow hydrocooler.

$$C = P \cdot T / 60$$

where

C = conveyor capacity (metric tons or tons, consistent with P)

P = weight of produce cooled per hour (metric tons per hour or tons per hour)

T = cooling time (minutes).

For an example in customary units, figure 1.40 shows that fruit 2 inches in diameter, like a plum, has a half-cooling time of about 7 minutes and can be seven-eighths cooled in about 21 minutes. Figure 1.43 shows that with this cooling time and a throughput rate of 60 tons per hour, the cooler must be large enough to hold about 20 tons of produce.

The area of the cooler conveyor (the area under the water shower) can be calculated by dividing the conveyor capacity by the area needed per unit of produce weight.

$$A = C/B$$

where

A = cooler area (square meters or square feet, consistent with C and B)

B = produce weight per unit area (metric tons per square meter or tons per square foot)

C = conveyor capacity (metric tons or tons).

Plums typically weigh about 1,000 pounds in a typical 4-by-4-foot pallet bin and, if the bins are placed two high in the cooler, 1 ton occupies 16 square feet. Therefore, the cooler needs a cooling area of 20 tons divided by 1/16 ton per square foot, equaling 320 square feet.

Conveyor length depends on selecting a convenient conveyor width. For example, produce in bins requires a conveyor 4 feet wide for a single-bin design or a conveyor 8 feet wide to accommodate two bins across.

$$L = A/W$$

where

A = cooler area (square meters or square feet, consistent with C and B)

L = conveyor length (meters or feet)

W = conveyor width (meters or feet).

For the plum example, a cooler two bins wide will have a length of 40 feet.

Conveyor speed can be calculated with the following formula:

$$V = L/T$$

where

L = conveyor length (meters or feet)

T = cooling time (minutes)

V = conveyor speed (meters or feet per minute, consistent with L).

The two bin–wide conveyor for the plum cooler would need to travel at a speed of 2 feet per minute.

Water pump capacity can be calculated by selecting a water flowrate based on the earlier discussion and multiplying this rate by the conveyor area. The pressure that the pump operates against is called head, measured in water-column height. It is the sum of the water lift plus friction losses in the piping. Motor size for the pump should be determined based on performance data supplied by the pump manufacturer. It can be approximated with the following formulas. In SI units,

$$M = (Q \cdot p)/\varepsilon$$

where

M = motor power (kW)

Q = water flowrate ($m^3 \cdot sec^{-1}$)

p = pressure (kPa)

ε = pump efficiency (usually between 0.50 and 0.70).

In customary units,

$$M = (Q \cdot H)/(\varepsilon \cdot 3960)$$

where

M = motor power (horsepower)

Q = water-flow rate (gpm)

H = static head (ft)

ε = pump efficiency (usually between 0.50 and 0.70).

Refrigeration

Most hydrocoolers use mechanical refrigeration. When a cooler is operated in a refrigerated room, as is the case for fresh-cut vegetables, the heat load will come entirely from the produce and the water pump. Hydrocoolers operated outdoors have additional conduction and infiltration heat loads. Tests on uninsulated hydrocoolers operated in the open sun have shown that these loads average 50 percent of the produce load (Thompson and Chen 1989). Conduction and infiltration loads can be measured fairly simply in an existing hydrocooler (under the "Cooling calculations" heading below, see the section on measuring the exterior heat gain of a hydrocooler). Batch coolers are often operated without doors—in fact, without any sort of air barrier on the openings where produce is loaded and unloaded. Continuous-flow coolers may have plastic flap covers on the entrance and exit openings, but they do not completely seal the cooler from infiltration. Hydrocoolers are difficult to insulate because splashing water wets most types of insulation. The refrigeration load for hydrocoolers with large water reservoirs is sometimes based on the amount of refrigeration required to quickly cool the water after the reservoir has been flushed and refilled. See the "Cooling calculations" section below for details of this calculation. Hydrocooler refrigeration systems are often sized based on the experience of the cooler designer rather than heat flow calculations.

Ice is used in some small coolers for crops with short harvest seasons. This approach is expensive and, because large blocks of ice lack adequate surface area to keep water at sufficiently low temperatures, a lot of labor is required to break the ice into fist-sized pieces and load it into the cooler.

Refrigeration evaporator coils are traditionally installed in the water reservoir below the produce. Placing the evaporator above the distribution pan allows faster water flow past the coil, produces a better heat exchange rate, and allows for use of a smaller, less expensive coil. Top-mounted coils also allow the reservoir to be smaller, reducing both energy use and the amount of water needed to fill the hydrocooler. If evaporator coils are not placed in the water reservoir, the reservoir only needs to be large enough to accommodate the pump intake and screens—and, during operation, to accommodate the volume of extra water needed to handle the water drawdown that is caused by filling the pipes, distribution pan, and packages.

Reducing electricity costs

Many electric utilities encourage their customers to shift electricity use to the night and morning hours by charging more for electricity use in the afternoon and early evening hours. Unfortunately, produce usually requires initial cooling during the day. Daytime electricity demand can be reduced by using an ice-accumulator refrigeration system. A refrigeration unit of moderate capacity freezes water on evaporator coils that are built into a water tank. The system can operate 24 hours a day, or only during the off-peak electrical rate period, making ice that melts during the on-peak period, when refrigeration demand is great. The overall economics of this system is favorable if the electricity load–shifting equipment is a part of a new installation. The cost of ice-building equipment is often equal to the cost savings available through installation of a smaller-capacity compressor and condenser. When water freezes around the coils, dirt tends to concentrate in the remaining unfrozen water. When there is a maximum ice load on the coils, however, much of this accumulated dirt can be drained with a minimal loss of refrigeration.

An energy accounting of two stone-fruit hydrocoolers (Thompson and Chen 1989) showed that 53 percent of the refrigeration energy used was for produce cooling. The rest was used for removing heat conducted across exterior surfaces of the cooler and pipes and through air infiltration (27%); removing heat added by the water pumps (10%); cooling water before start-up (8%); and cooling well water added to replace leakage (2%). Hydrocooler energy use can be reduced by

- insulating the cooler and piping, restricting air infiltration, and either installing a shade over the cooler or installing the cooler inside an insulated cold room (an uninsulated hydrocooler can provide most if not all of the refrigeration needed for the room enclosure)
- minimizing head pressure and minimizing the amount of water pumped
- using a shower pan instead of pressurized nozzles to distribute water
- reducing the water reservoir volume to lower the amount of water cooled at start-up
- designing the cooler exit to catch most of the water that drains from the produce as it leaves the cooler
- not cooling culled fruit (by cooling after the culling process in the packinghouse)
- operating at maximum capacity during the day (this minimizes per-unit energy use associated with pumps and with removing conduction and infiltration heat; these energy uses depend on the amount of time the cooler is in operation, not on the amount of produce cooled)

Measuring the exterior heat gain of a hydrocooler

Heat gain from conduction and air infiltration, along with heat from the water pump, can add up to a significant proportion of the total load on the refrigeration system. The magnitude of these heat sources can be estimated by cooling the water reservoir to operating temperature. Then, with the water pump operating and the refrigeration system off, measure the rate of temperature rise in the water in the reservoir. Water temperature should be measured every few minutes for about 15 minutes. The following formula relates the average water warming rate to the refrigeration capacity needed to remove the heat contributed by sources other than produce cooling and water replenishment:

$$R = \{[(m_c \cdot c_c + m_w \cdot c_w) \cdot DT] - Qp\}/k$$

where

R = refrigeration capacity needed to remove external heat gain (kW or tons)

m_c = weight of the empty cooler; includes only parts that are cooled (kg or lb)

c_c = specific heat of the cooler (assuming the cooler is made entirely of steel, its specific heat equals 0.46 $kJ \cdot kg^{-1} \cdot {}^\circ C^{-1}$ or 0.11 Btu/lb/°F)

m_w = weight of water in the cooler (kg or lb); 1 m^3 of water weighs 1,000 kg or 1 ft^3 of water weighs 62.4 lb

c_w = specific heat of water (4.19 $kJ \cdot kg^{-1} \cdot {}^\circ C^{-1}$ or 1 Btu/lb/°F)

DT = average rate of water temperature increase ($^\circ C \cdot hr^{-1}$ or °F/hr)

Qp = heat (in British thermal units per hour, or Btuh) added by water pump (assume 90% of the motor energy use ends up as heat in the water and 1 hp = 2,540 Btuh; for example, a 30-hp pump motor adds $30 \times 2{,}540 \times 0.90 =$ 68,580 Btuh; a fully loaded 22-kW motor adds 19.8 kW of heat)

k = 3,600 $s \cdot hr^{-1}$ or 12,000 Btuh/ton.

Following this procedure will estimate the exterior heat gain at the average ambient conditions during the test. Heat gain will vary with air temperature and humidity, wind conditions, sun location, and cloud cover. For example, maximum heat gain can be determined by conducting a test at solar noon, on a hot, clear day with a strong wind.

Hydrocooler operation and troubleshooting

Water should be changed daily unless the produce is particularly clean when entering the cooler. The level of chemicals used to control waterborne pathogens should be checked at least daily. If a lot of trash or dirt accompanies incoming produce, the concentration of chemical agents may drop rapidly. In such cases, concentrations should be checked several times a day.

Proper operation of a hydrocooler requires regular measurement of produce temperature. The cooler operator should measure and record the incoming and outgoing temperature of each lot of produce. Produce from various locations in the cooler should be sampled. With some experience, an operator will learn where in a particular cooler the warmest produce is likely to be. If outgoing fruit temperature rises above the desired range, cooler operation needs to be checked carefully. (See the "Measuring produce temperatures" section in chapter 2 for more details on produce temperature measurement.)

If produce temperature does not meet standards, the cooler may be poorly maintained or incorrectly operated. The following sections describe how to diagnose the causes of poor hydrocooler performance.

Troubleshooting shower-type hydrocoolers

In shower hydrocoolers, with the refrigeration evaporator installed above the water distribution pan or between the reservoir and the pan, clues about reasons for poor cooling performance can be found in the water temperature in the water distribution pan above the produce—as well as the water temperature in the reservoir under the produce.

Symptom 1: Water temperature increases in the distribution pan and in the reservoir during a cooling cycle or increases as the air temperature rises during the day.

Cause: Inadequate refrigeration capacity.

To solve the problem, reduce external heat gain by insulating the cooler, protecting it from the sun, installing plastic flap curtains or roll-down doors to restrict air infiltration, or installing the cooler in a cold room. Heat entering the cooler from air infiltration and conduction across the walls can equal 25 to 50 percent of the refrigeration requirements of a hydrocooler installed in the open.

Other options are to

- reduce incoming produce temperature through initial cooling with high-quality well water (water should be used only once, then used for irrigation; this initial cooling also cleans the produce, reducing trash build-up and soil accumulation in the cooler)
- check refrigeration maintenance
- increase refrigeration capacity (capacity can be increased in many ways, depending on the design of a particular system; in ice-operated coolers, capacity problems can be solved by adding more ice or breaking the ice into smaller pieces)
- put less produce on a batch cooler; on a continuous-flow cooler, slow the conveyor

Symptom 2: Water temperature is warmer than normal in the reservoir during a cooling cycle but remains cold in the distribution pan. Produce is not cooled to desired temperature.

Cause: Inadequate water flow through the distribution pan. A continuous-flow cooler with a shallow produce depth (as with cherries) requires about 4.8 to 6.8 $l \cdot sec^{-1} \cdot m^{-2}$ (7–10 gpm/ft^2); a batch or continuous-flow cooler for a deep produce bed (pallet loads of celery) requires 13.6 to 17.0 $l \cdot sec^{-1} \cdot m^{-2}$ (20–25 gpm/ft^2).

To solve the problem,

- check water pump
- check trash screens
- check for debris plugging the shower pan holes
- install more water-flow capacity and perhaps increase the number of holes in the distribution pan
- put less produce on the cooler

Symptom 3: Water temperature remains cold in the reservoir and in the distribution pan but produce cools slowly or in a nonuniform way.

Cause: Inadequate water flow through packaged produce.

To solve the problem,

- increase top and bottom package vent area
- ensure that package vents align if packages are stacked in the cooler
- consider using a different packing method for leafy produce, which may shed water like shingles on a roof
- ensure that water is supplied uniformly to the distribution pan

Cause: Attempt by operator to cool produce faster than heat can be removed from produce interior (see fig. 1.40 for typical cooling times).

To solve the problem,

- leave produce in the cooler longer
- reduce incoming produce temperature by using shades in receiving area, harvesting earlier in the day or at night, and quickly transporting produce to the cooler after harvest

Troubleshooting immersion hydrocoolers

Water temperature is the first thing to check when troubleshooting immersion coolers. Common problems are poor water flow and failure to leave the produce in the cooler long enough. If the produce floats, it is better suited to a shower cooler.

Symptom 1: Water temperature is cold enough (0–0.5°C [32–33°F]) for non–chilling sensitive produce, but produce is not adequately cooled.

Cause: Water agitation or water movement past the produce is not adequate. To solve the problem,

- increase pump capacity to increase water speed past the produce; relative speed of the water past the produce should be at least 0.075 $m \cdot sec^{-1}$ (15 fpm)
- add a water agitation system
- reduce incoming produce temperature

Cause: Trying to cool faster than heat can be removed from produce interior. To solve the problem,

- leave produce in the cooler longer
- reduce incoming produce temperature by

using shades in receiving area, harvesting earlier in the day or at night, and quickly transporting produce to the cooler after harvest

Symptom 2: Water temperature is not low enough; water should be at least 0.5° to 1.0°C (1–2°F) colder than the desired temperature of the produce when it exits the cooler.

Cause: Inadequate water-cooling capacity.

To solve the problem,

- put less produce through the cooler
- reduce external heat gain (most immersion coolers are installed as part of a packinghouse, so they are usually inside a building; if they are outside, heat gain can be reduced by insulating the cooler, protecting it from the sun, installing plastic flap curtains, or installing the cooler in a cold room)
- replace an air bubbler with another type of water agitation system, or design the air-handling system to recirculate air from the bubbler
- reduce incoming produce temperature by cooling it initially with high-quality well water; water should be used only once for cooling, and then used for irrigation
- check refrigeration maintenance
- increase refrigeration capacity; this can be done in many ways, depending on the design of a particular system

Vacuum cooling

Leafy green vegetables are generally vacuum-cooled (figs. 1.44 and 1.45): The produce is placed in an airtight chamber and the atmospheric pressure is reduced. This reduces the boiling temperature of water to field temperatures, and boiling conditions cause rapid water vaporization. The heat of water vaporization comes from the produce, causing it to cool.

Air is pumped out with one or more rotary-vane vacuum pumps, and water vapor is condensed on refrigerated surfaces located inside the cooling chamber (fig. 1.46). A cooler for 640 to 800 cartons of lettuce (with 23–27 kg [51–59 lb] of lettuce per carton) has 224 kilowatts (300 hp) of vacuum pumps and 370 kilowatts (500 hp) of refrigeration compressors (Thompson et al. 1987). Recently built vacuum coolers hold 24 pallets. Early cooler designs used a steam-jet pump to remove both air and water vapor.

The cooling process

Vacuum cooling has three distinct phases (fig. 1.47). The first phase, pressure reduction, begins when the cooler doors are closed and

Figure 1.44. Vacuum cooler with a capacity of twelve pallets (320 cartons).

Figure 1.45. Portable refrigeration and vacuum equipment for a vacuum cooler.

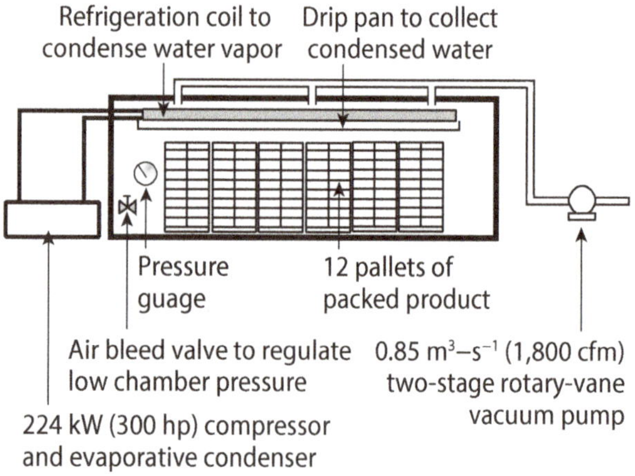

Figure 1.46. Key components of a vacuum cooler with a capacity of twelve pallets.

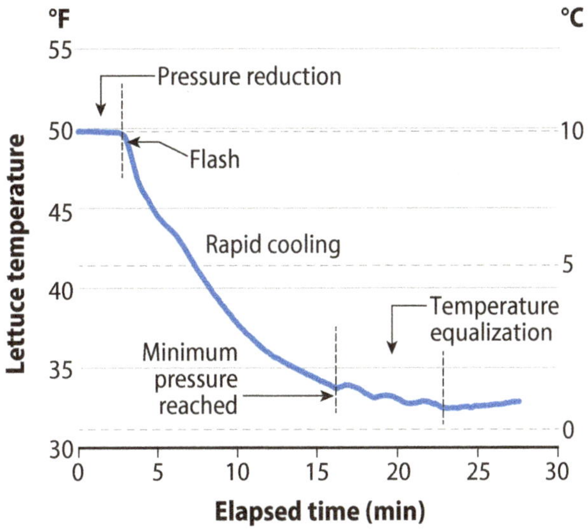

Figure 1.47. Romaine lettuce temperature in a typical vacuum-cooling cycle.

sealed. Vacuum pumps reduce the pressure in the cooler. During the pressure reduction phase, produce cools slowly. The cooling is associated with increased evaporation, likely caused by the declining partial pressure of water vapor in the cooler. After about 8 to 10 minutes the pressure has been reduced to levels that cause rapid produce cooling, and the second phase begins. The initiation of rapid cooling, called the "flash," occurs when the air pressure in the cooler drops to the saturation vapor pressure associated with the temperature of the produce. At this pressure the produce is under pressure conditions that cause water to boil and cause the rapid release of water vapor. The relationship between saturation vapor pressure and temperature is described by the following equation, assuming the water in the produce behaves like pure water (Buck 1981) (table 1.9):

$$e_w = 0.61121 \cdot EXP[(17.502 \cdot T)/(240.97+T)]$$

where

e_w = saturation vapor pressure (kPa absolute)

T = temperature (°C).

Based on this equation, lettuce placed in the vacuum cooler at 9°C (48°F) will begin rapid cooling when the cooler pressure drops to 1.148 kilopascals (8.609 mmHg) (see table 1.9).

During the second phase, rapid cooling, the produce cools rapidly until the cooler pressure drops to 0.61 kilopascals (4.58 mmHg), where water boils at 0°C (32°F). The pressure then is held constant. In commercial coolers, the time from flash to the minimum pressure is about 15

Table 1.9. Relationship between produce temperature and saturation vapor pressure (absolute) based on the Buck equation

Temperature		Saturation pressure		Temperature		Saturation pressure	
°C	°F	kPa	mmHg	°C	°F	kPa	mmHg
0	32.0	0.611	4.584	11	51.8	1.312	9.843
1	33.8	0.657	4.928	12	53.6	1.402	10.516
2	35.6	0.706	5.295	13	55.4	1.497	11.230
3	37.4	0.758	5.685	14	57.2	1.598	11.985
4	39.2	0.813	6.101	15	59.0	1.705	12.785
5	41.0	0.872	6.543	16	60.8	1.817	13.632
6	42.8	0.935	7.014	17	62.6	1.937	14.527
7	44.6	1.002	7.514	18	64.4	2.063	15.474
8	46.4	1.073	8.045	19	66.2	2.196	16.474
9	48.2	1.148	8.609	20	68.0	2.337	17.531
10	50.0	1.228	9.208	25	77.0	3.167	23.755

to 25 minutes for lettuce. Produce with warmer initial temperature takes more time to reach minimum pressure than does cooler produce. The warmer produce increases the amount of water vapor released from the produce, resulting in higher suction pressure in the condenser coil (evaporator) and less-efficient removal of water vapor. This vapor reduces the capacity of the vacuum pump (or pumps) to remove air from the vacuum tube and slows cooling.

At the beginning of the third phase—the temperature equalization phase—the fastest-cooling parts of the produce have reached 0°C (32°F). These parts of the produce are usually leaves, which rapidly release water vapor through stomata, lenticels, and other surface openings. The solid portion of the produce has limited ability to release water vapor—for example, the stem of a lettuce head—and cools more slowly (Rennie et al. 2001). The solid portion probably cools through slow release of water vapor and heat conduction to the faster-cooling portions of the produce. The solid portions of the produce finish cooling during the temperature equalization period. During this last part of the cooling cycle, the cooler pressure is often controlled at 0.61 kilopascals (4.48 mmHg) with a valve that allows small amounts of air to enter the cooler. The pressure is usually controlled with an on-off controller that sometimes causes an oscillating pressure (sometimes called "bounce") and temperature (see fig. 1.47). As this process continues, average produce temperature and temperature variability of the produce decrease. The length of time the produce must be held at 0.61 kilopascals (4.58 mmHg) depends on the type and initial temperature of the produce. Iceberg lettuce leaves require just a few minutes; romaine lettuce, with a solid stem and leaf ribs, requires 10 to 15 minutes; and cauliflower requires 2.5 to 4 hours (fig. 1.48). Warmer initial produce temperatures increase the length of the temperature equalization period compared with cooler temperatures.

Produce items with low surface-to-volume ratios, like fruit, or with few stomata or lenticels, like broccoli and cauliflower, must be

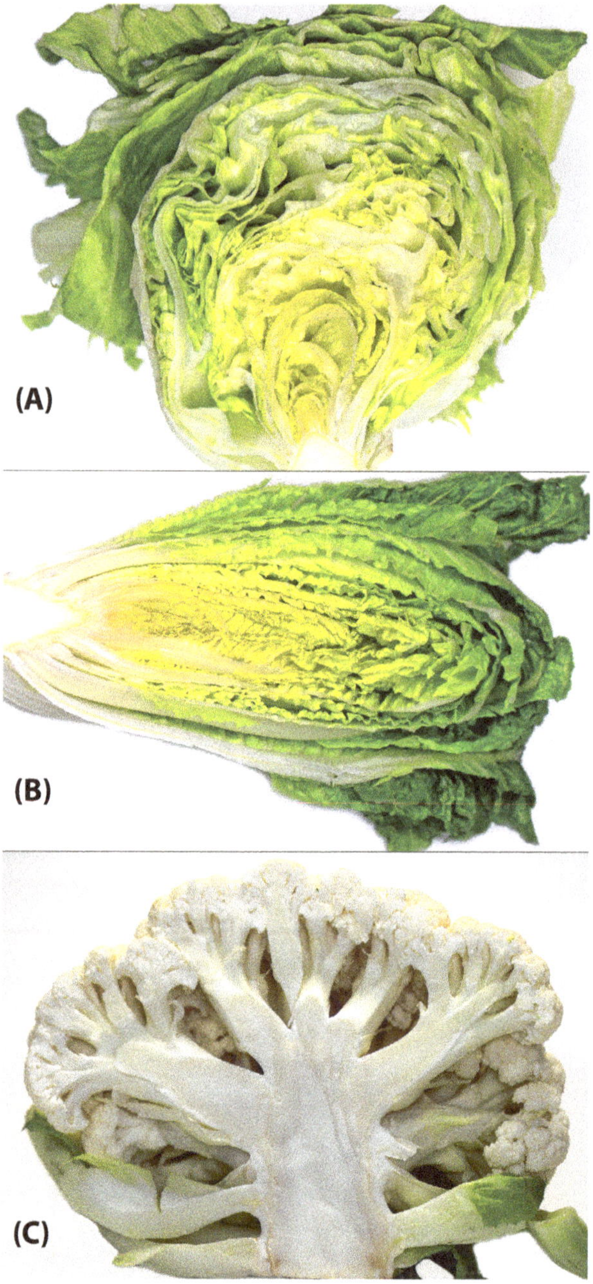

Figure 1.48. Cross sections of iceberg lettuce (A), romaine lettuce (B), and cauliflower (C). *Photos:* Evett Kilmartin.

held at 0.61 kilopascals (4.58 mmHg) for many hours, so other methods are used to cool them.

Vacuum cooling causes water loss equal to about 1 percent of the produce weight for every 5.0 to 5.6°C (9–10°F) of cooling (Guillou 1958; Barger 1963). Two to 4 percent weight loss is common, and this level can cause noticeable wilting of some leafy vegetables. Spraying water on produce before it enters the vacuum cooler allows wilt-sensitive vegetables to be cooled with no wilting. Much of the cooling effect comes from evaporation of the added water rather than produce moisture. Water-spray vacuum coolers are designed to spray water on the produce during the cooling cycle. They are simpler to operate than processes that use a separate spray step ahead of cooling. Evaporation of surface water also causes faster cooling for produce with low transpiration rates. Vegetables such as green onions, leaf lettuce, celery, and spinach are often cooled in water-spray vacuum coolers.

Rapid release of air into the vacuum chamber at the end of water-spray cooling can cause leaves to appear water-soaked. The damage is apparently caused when surface water is forced into the tissue by the rapid pressure rise around it. Damage is eliminated by releasing the vacuum over several minutes.

Plastic bags or wraps around a vegetable can slow water-vapor release and can slow cooling (Harvey 1963). Tiny holes in the plastic will allow enough water vapor to escape for rapid cooling while still reducing water loss during subsequent marketing (Cheyney et al. 1979). Plastic-film box liners will not slow vacuum cooling if the liners are folded over the top of the produce without sealing them.

Monitoring

There are several methods for estimating or directly measuring produce temperature in the vacuum-cooling process. A high-resolution and well-calibrated gauge that measures absolute pressure in the chamber gives a direct indication of the water-boiling temperature in the chamber and provides a good estimate of the temperature of the coldest parts of the produce during cooling. A wet-bulb thermometer (which measures the boiling temperature of water in a vacuum) can be used as an indicator of minimum produce temperature for rapidly cooled vegetables. Temperature probes can be inserted into the produce and observed during cooling with a radio connection to an exterior readout device. However, these units are not commonly used because operators often forget to retrieve the temperature probes after the produce is removed from the cooler. Pulp temperature of produce leaving the cooler should be routinely checked to ensure produce is cooled to an acceptably low temperature without freezing.

Equipment

Vacuum cooling equipment is quite costly. As of 2017, a new cooler with capacity for 24 pallets cost about $1.25 million. To be economically feasible, the equipment must be used consistently throughout the year. Most vacuum coolers are trailer-mounted so they can be moved as production areas change during the year. Coolers have been built small enough for cooling a single pallet and large enough for cooling 60 pallets per cycle.

Vacuum-pump capacity is often designed to drop the cooler pressure from atmospheric pressure—101 kilopascals (760 mmHg)—to the initiation of cooling at about 2.67 kilopascals (20 mmHg) in 8 to 10 minutes.

Refrigeration

The highly variable rate of cooling in vacuum coolers produces great fluctuation in refrigeration demand. In a test with the cooler described in figure 1.46, electricity demand peaked initially at 350 kilowatts and dropped to 170 kilowatts at the end of the cooling cycle (fig. 1.49).

The refrigeration equipment is sized to handle the high water-vapor load and rapid cooling at the beginning of cooling. Tests with commercial vacuum coolers show cooling rates of 5.6°

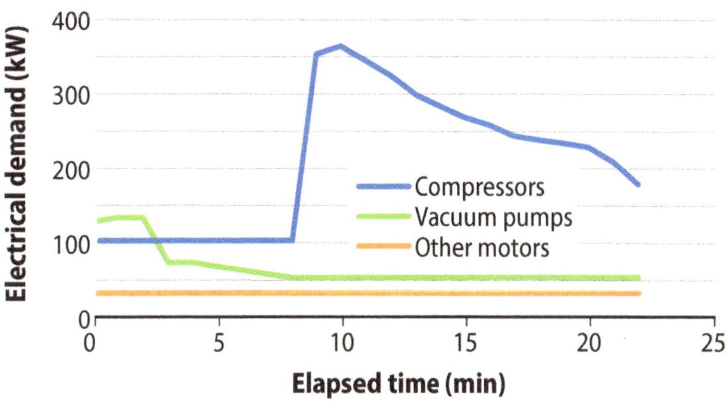

Figure 1.49. Electricity demand (kW) for a load of lettuce cooled by vacuum cooler described in figure 1.46. *Source:* Thompson et al. 1987.

to 6.1°C·min^{-1} (10–11°F/min) during the first minute of cooling for rapid-cooling items like very immature lettuce leaves and warm romaine lettuce (25°C or 77°F). Iceberg lettuce and romaine lettuce with an initial temperature near 15°C (59°F) have a cooling rate of 2.8°C·min^{-1} (5°F/min) in the first minute. The high rates of moisture release last for only the first few minutes of cooling and, after 10 minutes, the cooling rate drops to about 0.8°C·min^{-1} (1.5 °F/min) for most commodities. If a refrigeration system is undersized, it will result in significant increases in cooling time. For example, if a system is limited to cooling at a rate of only 1.1°C·min^{-1} (2°F/min), total cooling time will increase by 7 to 8 minutes for fast-cooling vegetables. This adds about 25 percent to the total time for a typical lettuce load and significantly reduces the daily capacity of the cooler. Produce cooling is the only heat load in a vacuum cooler. Refrigeration capacity for cooling the spray water is slightly greater because the boxes in a cooler accommodating twenty-four pallets carry out about 380 liters (100 gal) of water that must also be cooled.

When multiple vacuum coolers are on a common refrigeration system, the best practice is to time the start of each cycle so that the high refrigeration demand of a new cycle does not slow cooling of the prior cycle. This strategy yields the lowest evaporating temperature in the condensing coil, maximizing the temperature differential between the produce being cooled and the liquid refrigerant that cools the condensing coil.

Reducing electricity costs

Vacuum coolers are very energy-efficient compared to other types of coolers. They remove heat only from the produce, whereas the refrigeration system in a forced-air cooler also removes heat from lift trucks, lights, fans, walls, and air infiltration (Thompson and Chen 1988). Electricity use can be reduced in vacuum coolers by reducing vacuum-pump capacity after rapid cooling begins in a cycle; filling the retort with as much produce as possible; and shutting down refrigeration compressors or switching their capacity to other coolers when refrigeration is not needed during the cooling cycle. Do not use vacuum coolers to cool produce that requires long cooling times (Thompson et al. 1987). In one study (Pacific Gas and Electric 2008), electricity use was cut by 25 percent by installing motor-speed control devices to reduce the capacity of screw compressors, vacuum pumps, and condenser fans when their full output was not needed.

Package icing

Packing finely crushed or flaked ice with perishables is a traditional method of cooling, and it is particularly valuable for produce shipped in nonrefrigerated vehicles. Ice is often packed in flower boxes because they are frequently delivered by nonrefrigerated public or commercial mail services. Ice also maintains a high humidity level in the box and minimizes moisture loss once the produce has cooled to the temperature of the ice.

Fast cooling with ice requires that the ice be in contact with the produce. Just placing ice on the top of a carton of produce (called top icing) does not cause fast cooling. Injecting a slurry of ice and water, a process called liquid icing or ice-injection cooling, is an effective way to distribute ice in boxes. Water is mixed with ice particles 0.1 to 0.5 millimeters in size, creating a slurry with a consistency similar to partly melting snow. For rapid cooling, the slurry usually contains 6 to 20 percent ice. Sodium chloride is added to the slurry to lower the water freezing point to about –1.1°C (30°F). The water-ice slurry is pumped into the hand holes of waxed, corrugated, fiberboard cartons. The slurry flows throughout the box, ice fills the voids, and the water drains through holes in the bottom of the box. Initially, the cooling process removes sensible heat from the produce, similar to what happens in hydrocooling. The residual ice, about 10 kilograms (22 lb) per box, continues the cooling process after slurry injection is completed and provides a refrigeration effect during subsequent transportation. The slurry can be injected with a manually operated nozzle (fig. 1.50) or an automated system (fig. 1.51). The automated system takes only 3 or 4 minutes to uniformly ice a pallet load of boxes.

Flaked or crushed ice can be manufactured on-site with several types of commercially available equipment. It is usually manufactured during periods with off-peak electrical rates and stored in an ice bunker, allowing the refrigeration system to use inexpensive electricity. If the cooling season is short and ice use only amounts to a few tons per day, it is often cheaper to buy block ice and transport it to an on-site crusher. Liquid-ice equipment can be made portable, and some companies lease the equipment and move it to the cooling site as needed.

Ice has a cooling capacity of 335 kJ·kg^{-1} (144 Btu/lb) and has the capacity to cool three times its weight in produce, assuming about 28°C (50°F) of temperature drop in the produce. However, commercial ice-injection systems require much more ice than is needed just for produce cooling. Significant amounts of ice melts in the typically uninsulated ice injection equipment. Extra ice may be added to the box to maintain produce temperature in transport and have some ice present in the box when it arrives at its destination. The high ice requirement makes liquid icing an energy-inefficient and expensive cooling method.

Manufactured ice presents a number of other disadvantages. The weight of the ice reduces the amount of produce that can be transported in a highway vehicle. Ice requires the extra expense of using a water-resistant container. In a mixed load, the water produced by the melting ice can damage neighboring containers that are not water-resistant, and the water is a safety hazard in warehouses. For packages shipped by air freight, release of free water is not allowed.

Some of these disadvantages can be eliminated if the ice is not used for cooling but only for maintaining temperature in transit. Flowers are often packed with plastic packs filled with gelled water that do not leak when pierced. The gel-ice pack is placed in the box during packing and the box can then be forced air–cooled. The box vents are closed after cooling.

Figure 1.50. Manually operated ice-injection system.

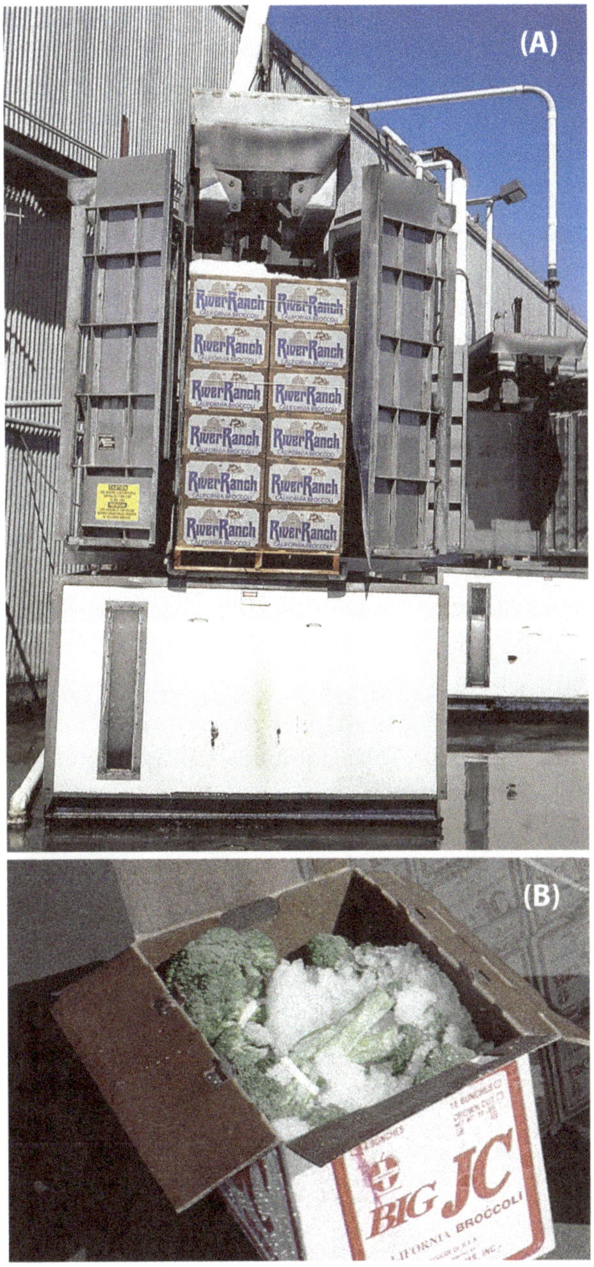

Figure 1.51. Automated ice injector for palletized produce (A) and waxed broccoli box filled with crushed ice (B).

Transport cooling

Typical refrigerated highway trailers have limited airflow and limited refrigeration capacity but can cool low-density items like cut flowers in transit. Cooling times of 24 hours are common in loads of cut flowers. A full trailer load of cut flowers weighs only about 9,070 kilograms (20,000 lb), but a load of most other perishable items weighs over 18,140 kilograms (40,000 lb) and cannot be successfully cooled in highway transport. In fact, most loads of fruit and vegetables shipped in the United States warm slightly even in refrigerated transport. Fruit and vegetables should be cooled to their desired transit temperature before being loaded in highway trailers.

Refrigerated marine containers, cargo ships, and railcars can achieve some cooling during transport, but cooling is slow. Containers and refrigerated ships usually have bottom-delivery airflow and refrigerated air is supplied to the produce through a floor plenum (fig. 1.52). Effective cooling is obtained when the floor is completely covered with produce and the fork-lift openings in pallets near the rear doors have been blocked. This forces the refrigerated air to flow vertically through the boxes. Uncovered areas of the floor allow poor air distribution in the plenum and allow air to flow around boxes rather than through them. Boxes must have top and bottom vents to allow vertical airflow through the packages, and vents must align between layers of containers. Inner packaging materials must not block vertical airflow.

Flaked or crushed ice or liquid ice (ice-water slurry) is sometimes distributed on top of warm loads in refrigerated railcars. The ice supplements mechanical refrigeration during initial cooling of a warm load. Crushed ice solidifies, however, forming a blanket that may block air circulation through the load and cause slow cooling. Interference with a vehicle's air circulation can be minimized by top-icing in windrows along the length of the load, ensuring that no ice blocks the space between the load and the side wall. This allows air to enter the load freely through the open areas between the windrows. Cantaloupes in wooden crates have been satisfactorily cooled in less than 24 hours by combining this method of icing with operation of fans and mechanical refrigeration in railcars. Portable auxiliary fans cannot increase airflow through the load sufficiently to significantly speed cooling under these conditions (Kasmire and Parsons 1971).

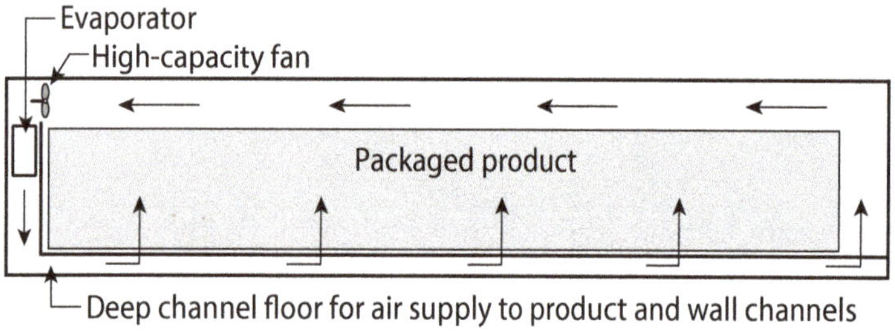

Figure 1.52. Airflow in a bottom-air delivery system for marine containers. A similar airflow is used in refrigerated breakbulk ships.

Cooling methods for developing countries

A number of cooling options are available for small-scale producers in developing countries (Kitinoja and Thompson 2010). In forced-air evaporative cooling, the air is cooled with an evaporative cooler instead of with mechanical refrigeration. If designed and operated correctly, an evaporative cooler produces air a few degrees above the outside wet-bulb temperature, and the cooler air is above 90 percent relative humidity. In most areas of California, produce temperatures of 16° to 21°C (60–70°F) can be achieved. The evaporative cooling pad is continuously supplied with outside air and the air leaving the packages is vented to the outside, away from the air intake. An example of a forced-air evaporative cooler is shown in figure 1.53. This cooling method may be adequate for produce that is best held at moderate temperatures, such as tomatoes, or for produce that is marketed quickly after harvest. In most cases, growers can build their own forced-air evaporative coolers. These units are more energy-efficient than mechanical refrigeration. Passive evaporative cooling technology, like wetted, unglazed clay pots and masonry walls built with a center layer of wetted sand, have been used for centuries to provide relatively cool, humid storage conditions. They provide room-cooling conditions and are best suited for produce that has not been packaged. Their minimum air temperature is also limited to a few degrees above the prevailing ambient wet-bulb air temperature.

Lower air temperatures can be produced using a special controller (see, for example, storeitcold.com) installed on a room air conditioner. The controller allows retail air conditioners to maintain cold-room air temperature as low as about 2°C (36°F). The controller and air conditioner can be purchased for about 90 percent less than equivalently sized commercial refrigeration equipment. The controller causes very little condensation on the refrigeration coil and the unit is likely to produce relatively high humidity in the cold room.

Electricity may not be available on a dependable basis in many developing countries. Locally manufactured ice blocks are commonly available in many areas and they can be used quite effectively to provide cold, humid conditions. This method was commonly used in the United States in rail transportation until World War II (fig. 1.54). The blocks are placed in a slotted, wall-mounted bunker and a fan draws air past the ice. Ice placed in produce boxes provides some temperature control in unrefrigerated handling.

Selecting a cooling method

Many commodities can be cooled with several types of coolers. Selecting the best type of cooler for a particular operation depends on several considerations, such as the mix of commodities handled, the length of the cooling season, the need for fast cooling, and so on. Table 1.10 compares the common cooling methods.

Forced-air cooling, the most adaptable cooling method, is widely used for fruit, fruit-type vegetables, and cut flowers (table 1.11). Most perishable commodities can be effectively cooled with this method. Its main disadvantage is water loss, which can be reduced by minimizing delay between harvest and the start of cooling. Delays cause moisture loss by allowing warm produce to be exposed to low ambient humidity. If the produce gains temperature in the delay, more moisture will be lost in cooling. Applying waxes, or packing produce in plastic bags, also reduces moisture loss when this cooling method is used.

The second-most commonly used cooling method is hydrocooling, which is used for root, stem, and flower vegetables; some tree fruit; and fresh-cut vegetables. Produce exposed to adequate water flow cools quickly with no moisture loss. In some cases, wilted produce actually gains some moisture. The main disadvantages of hydrocooling are that recirculated water must

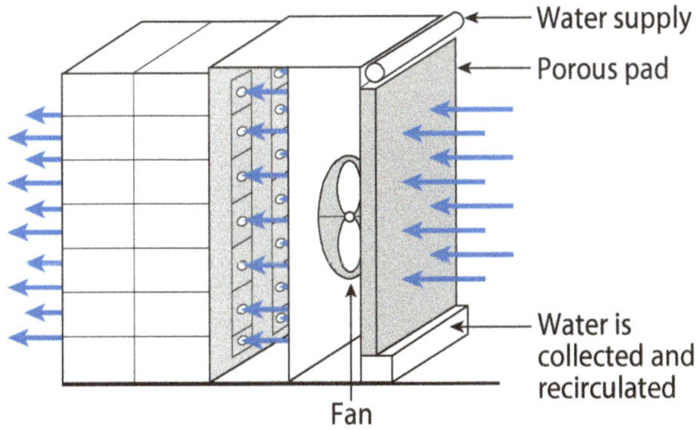

Figure 1.53. Cutaway view of an evaporative forced-air cooler.

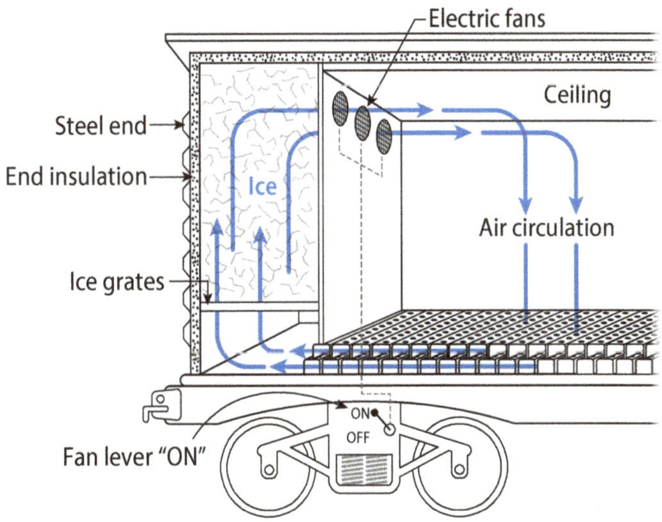

Figure 1.54. Cutaway drawing of an ice-refrigerated railcar.

Table 1.10. Comparison of product effects and cost for six common cooling methods

Cooler characteristic	Forced air	Hydro	Vacuum	Water-spray vacuum	Ice	Room
typical cooling time (hr)	1–10	0.1–1	0.3–2	0.3–2	0.1–0.3*	20–100
product moisture loss (%)	0.1–2.0	0–0.5	2.0–4.0	no data	no data	0.1–2.0
water contact with product	no	yes	no	yes	yes	no
potential for decay contamination	low	high	none	high	medium	low
water-resistant packaging needed	no	yes	no	yes	yes	no
capital cost	low	low	medium	medium	high	low†
energy efficiency	usually low	high	high	medium	low	low
portable	sometimes	rarely done	common	common	common	no
feasibility of continuous conveyor cooling	rarely done	yes	no	no	rarely done	no

*Top icing can take much longer.

†Low if product is also stored in cooler, as with apples; otherwise, long cooling times make room cooling an expensive system.

Table 1.11. Cooling methods suggested for horticultural commodities

Commodity	Size of operation		Remarks
	Large	Small	
fruit			
all berries, kiwifruit	FA	FA	—
grapes	FA	FA	Cooling facilities should be adaptable to SO_2 fumigation
mushrooms	FA, VC	FA	—
sweet corn	HC, VC, PI	HC, FA, PI	—
citrus	R, FA	R	—
stone fruit	FA, HC	FA	Apricots cannot be hydrocooled
pome fruit	FA, R, HC	R	—
subtropical	FA, HC, R	FA	—
tropical	FA, R	FA	—
leafy vegetables			
cabbage	VC, FA	FA	—
iceberg lettuce	VC	FA	—
kale, collards	VC, R, WVC	FA	—
leaf lettuce, spinach, endive, escarole, Chinese cabbage, bok choi, romaine	VC, FA, WVC, HC	FA	—

Continued on next page

Table 1.11. Cooling methods suggested for horticultural commodities, continued

Commodity	Size of operation		Remarks
	Large	Small	
underground vegetables			
with tops	HC, PI, FA	HC, FA	Carrots can be vacuum-cooled
topped	HC, PI	HC, PI, FA	—
Irish potatoes	R with evaporative coolers		With evaporative coolers, facilities should be adapted to curing
sweet potatoes	R	R	—
dry onions	R	R, FA	should be adapted to curing
garlic	R	R	—
stem and flower vegetables			
artichokes	HC, PI	FA, PI	—
asparagus	HC	HC	—
broccoli, Brussels sprouts	HC, FA, PI	FA, PI	—
cauliflower	FA, VC, HC	FA	—
celery, rhubarb	HC, WVC, VC	HC, FA	—
green onions, leeks	PI, HC, WVC	PI	—
pod vegetables			
beans	HC, FA	FA	—
peas	FA, PI, VC	FA, PI	—
fruit-type vegetables*			
cucumbers, eggplant, summer squash, okra	R, FA, FA–EC	FA, FA–EC	—
peppers	R, FA, FA–EC, VC	FA, FA–EC	—
tomatillos, tomatoes	R, FA, FA–EC	R, FA, FA–EC	—
winter squash	R	R	—
melons			
cantaloupes	HC, FA, PI	FA, FA–EC	—
honeydew, casabas, Crenshaw	FA, R	FA, FA–EC	—
watermelons	FA, HC	FA, R	—
fresh herbs			
not packaged	HC, FA	FA, R	can be easily damaged by water beating in HC
packaged	FA	FA, R	—

Continued on next page

Table 1.11. Cooling methods suggested for horticultural commodities, continued

Commodity	Size of operation		Remarks
	Large	Small	
cactus			
leaves (nopalitos)	R	FA	—
fruit (tunas, or prickly pears)	R	FA	—
ornamentals			
cut flowers	FA, R	FA	When packaged, only use FA
potted plants	R	R	—

KEY:
FA = forced-air cooling, FA–EC = forced-air evaporative cooling, HC = hydrocooling, PI = package icing, R = room cooling, VC = vacuum cooling, WVC = water-spray vacuum cooling

Note: "—" indicates that table cells are left intentionally blank.

*Fruit-type vegetables are chilling-sensitive, but at varying temperatures.

be carefully disinfected to prevent spread of human and produce decay pathogens; some produce, like most berries and mushrooms, do not tolerate water contact; and packaged produce must use plastic or water-resistant containers. Corrugated fiberboard boxes dipped in wax are significantly more expensive than untreated fiberboard boxes and are not easily recyclable.

Vacuum coolers and water-spray vacuum coolers are used mainly for vegetables that lose water easily, such as leaf vegetables. Short cooling times are needed to spread the high capital cost of this equipment over a sufficient amount of produce. Iceberg lettuce is usually vacuum-cooled; most other leaf vegetables are cooled with water-spray vacuum cooling. Water-spray vacuum cooling has all the disadvantages of hydrocooling. Vacuum coolers and water-spray vacuum coolers are often made portable so they can be used year-round as production areas change.

Ice cooling is limited to a few commodities that have been traditionally cooled with this method. Broccoli is often ice-cooled, and some cut flowers are packed with gel or crushed ice to maintain low temperatures in nonrefrigerated transport and storage conditions. Liquid-ice injection systems recirculate water and require the same water treatment as hydrocoolers. Hand-placed ice has less potential for decay spread, but melt water can spread microorganisms. Ice cooling is expensive, and water from melting ice is a nuisance and causes wet floors in transport vehicles and in cold storages. Ice increases gross package weight and usually reduces the amount of produce that can be shipped in a truck. The melt water can damage neighboring corrugated containers that are not designed for water contact.

Room cooling is best suited to situations in which the produce is individually exposed to cold air, like cut flowers stored in buckets or fresh-cut fruit and vegetables that are processed and packed in a cold room. Packaged produce cools slowly. Citrus fruit are still commonly room-cooled, and some controlled-atmosphere stored fruit are allowed to cool slowly so that their respiration can be used to lower oxygen and raise carbon dioxide concentrations in storage.

A cooling facility should always include cold-storage space. Produce exiting a cooler will quickly warm unless it is loaded directly into refrigerated transport vehicles or placed in a cold room. Rewarming wastes the benefits of cooling—and cooled produce left in a warm environment is subject to condensation, which may lead to development of disease on the produce. In many initial cooling operations, it is not feasible to leave produce in a cooler long enough for each piece to reach desired temperature. Final cooling can be achieved by holding the produce in a cold room designed for good room cooling. Faster final cooling can be achieved by setting up a portable fan and tarp unit to cool the produce with a forced-air system. Since the produce is close to the desired final temperature, little or no additional refrigeration capacity is needed to finish cooling, and low airflow rates can be used.

Cooling calculations

This section describes methods of calculating produce temperature and cooling time during a cooling process. Methods of calculating refrigeration load for forced air and hydrocooling are also described.

Cooling times and temperature

Exact calculations of time and temperature relationships in cooling produce involve conditions that are seldom known precisely—like airflow rate, initial produce temperature and its variability, and coolant temperature and its variability over time—and can be time-consuming to obtain. However, satisfactory estimates for design and operation of cooling facilities can be based on typical conditions and an assumption of logarithmic cooling, as depicted in figure 1.23 (Guillou 1960). (For more accurate, explicit mathematical models of cooling, see Stoecker 1988–1995 or ASHRAE 2022.) Assuming logarithmic cooling, cooling time is often described by the time needed to achieve seven-eighths cooling. This is the time required to cool produce by an amount equal to seven-eighths of the difference between initial produce temperature and the coolant temperature.

Table 1.12 is useful for making estimates of expected seven-eighths cooling time and temperature. For example, if after 50 minutes in a forced-air cooler the produce has cooled from 32.2° to 15.6°C (90° to 60°F), and the cooling air is held at −1.1°C (30°F), the produce is one-half cool and has dropped 16.7°C (30°F) in temperature. This represents one-half the difference between the initial produce temperature and the coolant temperature. Multiply 50 minutes by 3 to find an expected seven-eighths cooling time of 150 minutes. At seven-eighths cool, the produce will have cooled by an amount equal to 7/8 multiplied by 33.3°C (60°F) (the difference between the initial produce temperature and the coolant temperature). Subtracting seven-eighths the cooling amount from the initial produce temperature gives a seven-eighths cooling temperature of 3.1°C (37.5°F).

This approximation is a reasonably good estimate of the average temperature of a lot of produce. Average temperature is fairly easy to measure in hydrocoolers, where there is not much temperature difference between produce in various locations along the path of water flow. In forced-air coolers, however, the most exposed produce cools faster and the most protected produce cools more slowly than the calculation above indicates.

Seven-eighths cooling times are three times longer than half-cooling times. Seven-eighths cooling, like half-cooling, is the same for a given produce item in a given package regardless of the temperatures of the produce and coolant. Seven-eighths cooling times and temperatures are practical estimates of cooling in many commercial operations and are closer to commercial cooling times than half-cooling times. However, some operations now attempt to cool their produce to temperatures closer to their coolant temperature, and they use a fifteen-sixteenths cooling-time estimate. This is equal to four half-cooling periods. Seven-eighths cooling time

Table 1.12. Data for estimating expected seven-eighths cooling time and temperature

Ratio of product temperature drop during cooling divided by difference between initial product temperature and cooling medium	1:2	3:4	7:8	15:16	31:32
Multiply elapsed cooling time by the following to find seven-eighths cooling time	3	3/2	1	3/4	3/5

can be measured by recording produce temperature during cooling, but it is meaningful only when the coolant temperature is fairly constant.

Time to cool to any temperature can be calculated with the following formula:

$$T = -Z \ln(Y)/0.693$$

where

T = cooling time (min)

Z = half-cooling time (min)

(Half-cooling time = 1/3 × seven-eighths cooling time).

$0.693 = \ln(2)$

Y = unaccomplished temperature difference (dimensionless temperature ratio)

$$Y = (t - t_m)/(t_i - t_m)$$

(Y = 1 at the start of cooling and Y = 0 when the produce temperature equals the cooling media temperature).

where

t = desired produce temperature (°F or °C)

t_m = temperature of the cooling medium (°F or °C)

t_i = initial produce temperature (°F or °C).

Calculating momentary cooling rate

In designing cooling systems, it may be necessary to know the rate at which a batch of produce will be cooling at some specific time. Use

$$R = 2.08(t - t_m)/S$$

where

S = seven-eighths cooling time (h)

$(t - t_m)$ = produce-to-cooling media temperature difference (°F or °C)

R = momentary cooling rate (°F/h or °C·h^{-1})

2.08 = natural logarithm of 1/8.

This formula applies whether coolant temperature is constant or varying, and may also be used to estimate S from test measurements of R and $(t - t_m)$ (Guillou 1958).

Cooling coefficients

Cooling rates are sometimes designated by a cooling coefficient, or temperature reduction in degrees per hour divided by the average temperature difference between produce and coolant. For example, produce cooled by 4° (°C or °F) in 15 minutes will have cooled at an average rate of 16° per hour. If the average temperature difference between produce and coolant over the 15-minute period was 32°, the cooling coefficient was

$$C = 16^\circ \text{ per hour}/32^\circ = 0.5 \text{ degrees per hour per degree.}$$

Neglecting departures from idealized logarithmic cooling, the cooling coefficient is the same for any period of a cooling operation, or at any moment. Therefore, using C for cooling coefficient (degrees per hour-degree) (Guillou 1958),

$$R = C(t - t_m).$$

In practice, cooling coefficients may vary considerably as cooling progresses and should be measured over temperature differences that approximate actual temperature differences. Test results with variable coolant temperatures may be applied to operations with constant coolant temperatures, or vice versa, by the relation between cooling coefficient and seven-eighths cooling time:

$$R = 2.08/S$$

or

$$S = 2.08/C.$$

Refrigeration for forced-air coolers

The refrigeration capacity of an evaporator for an individual forced-air cooling unit is calculated using the same method as for cold storage. Heat inputs are summed from produce heat loss; fans and motors; heat loss from packing materials; air infiltration; heat conducted through cold-room exterior surfaces; heat input from lights, lift trucks, and people; and produce respiration. The largest heat load is from cooling the produce. Heat released from the produce is great at the beginning of cooling and drops rapidly during the process. It is not economically feasible to provide refrigeration capacity to remove the instantaneous initial heat load, but it is not necessary. Installing sufficient refrigeration capacity to handle 45 percent of the initial produce heat load only increases cooling time by 15 to 20 percent compared with installing enough capacity to remove the initial instantaneous heat load (Wade 1984). A simple and more conservative approach is to install enough refrigeration capacity to handle the average heat load during the first half-cooling period. This increases cooling time by less than 5 percent compared with refrigeration sized to handle the maximum produce heat load, while it reduces refrigeration capacity by about 25 percent. Refrigeration capacity is very dependent on initial produce temperature, and the design should accommodate the highest likely produce temperature.

In facilities with multiple cooling positions, the evaporator and other low-pressure equipment will be sized according to the above considerations. However, it is rare for all cooling positions to be turned on at the same time, so the compressor, condenser, and other high-pressure equipment can be smaller than the sum of the loads for all the low-pressure equipment.

Calculating the peak refrigeration demand requires knowing when during the day produce arrives and the temperature of incoming produce, and assuming an acceptable cooling rate. Table 1.13 shows an example calculation in customary units. The following assumptions are made in calculating refrigeration capacity. The example is in customary units.

- Weighted heat capacity of produce and packing materials, c_p = 0.9 Btu/lb-°F
- Cooling air temperature = 30°F
- Seven-eighths cooling time = 3 hours (half-cooling = 1 hr)
- Produce is removed from the cooler when it reaches seven-eighths cool
- Load is calculated based on average capacity needed for half cooling, using

$$R = (c_p \cdot \Delta T \cdot m)/k$$

where

R = refrigeration capacity (tons)

ΔT = temperature drop (°F) per hour based on a half-cooling time = 1 hr.

The produce cools by one-half the difference between its initial temperature and the temperature of the cooling air.

m = weight of produce (lb)

k = 12,000 Btuh/ton of refrigeration.

The peak refrigeration tonnage associated with produce cooling is 28.1 tons. Total refrigeration equals produce cooling load plus heat from other sources, such as fans, lift trucks, lights, air infiltration, people, and conduction through walls. As a rule of thumb, heat from miscellaneous sources equals 25 percent of the produce load. An extra 10 to 15 percent of the total heat load is added as a safety factor to account for facility use beyond assumptions—for example, starting two batches simultaneously or cooling unusually warm produce.

Many designers size refrigeration equipment based on their experience. They have observed equipment that has been used for a variety of commodities, a range of packaging systems, the extreme produce load conditions that occasionally occur at times of high harvest rates and high temperature conditions, and equipment operating under various levels of maintenance. This allows them to size a system that can handle the extreme load situations that may occur in commercial practice.

Refrigeration for hydrocoolers

Peak demand for batch coolers can be calculated using the method used for forced-air coolers. Continuous-flow hydrocooler refrigeration load can be calculated with the following formula:

$$R = (M \cdot c_p \cdot \Delta T)/k$$

where

R = refrigeration capacity (tons or kW)

ΔT = total temperature drop (°F or °C) of

Table 1.13. Example calculation of refrigeration load for a forced-air cooler

Time	Product load (lb)	Product temperature (°F)	Refrigeration load for each batch (tons): 10 a.m.	12 p.m.	2 p.m.	4 p.m.	6 p.m.	Total refrigeration load (tons)
10 a.m.	10,000	70	15.0	—	—	—	—	15.0
11 a.m.	—	—	7.5	—	—	—	—	7.5
12 p.m.	10,000	80	3.8	18.8	—	—	—	22.6
1 p.m.	—	—	—	9.4	—	—	—	9.4
2 p.m.	10,000	90	—	4.7	22.5	—	—	27.2
3 p.m.	—	—	—	—	11.3	—	—	11.3
4 p.m.	10,000	90	—	—	5.6	22.5	—	28.1
5 p.m.	—	—	—	—	—	11.3	—	11.3
6 p.m.	10,000	90	—	—	—	5.6	22.5	28.1
7 p.m.	—	—	—	—	—	—	11.3	11.3
8 p.m.	—	—	—	—	—	—	5.6	5.6

Note: "—" indicates that table cells are left intentionally blank.

produce during cooling

M = weight of produce entering cooler per time (lb/hr or kg·sec^{-1})

c_p = weighted average heat capacity of produce and packing material (typically about 3.77 kJ·kg^{-1} · °C^{-1} or 0.9 Btu/lb-°F)

k = 3,600 s·h^{-1} or 12,000 Btuh per ton of refrigeration.

Total refrigeration equals produce cooling load plus heat from miscellaneous sources—such as pumps, air infiltration, makeup water cooling, and conduction through walls. As a rule of thumb, heat from miscellaneous sources equals 25 percent of the produce load for a well-insulated cooler that is protected from wind and direct solar radiation and 50 percent for a poorly insulated cooler that is operated in the open. An extra 10 to 15 percent of the total heat load is added as a safety factor to account for facility use beyond assumptions. In some hydrocoolers, the maximum refrigeration demand may be for cooling the water before start-up. This is especially true if the water reservoir is large and the operator wants the water to cool quickly to operating temperature. This is important if water is changed during the day—for example, if it is changed at the noon break and the cooler must be ready for use after the break.

In some cases, the maximum refrigeration capacity for a hydrocooler is based on quickly cooling down a fresh batch of water after the hydrocooler has been drained. The formula below can be used to calculate refrigeration capacity for water cooldown, assuming water will be circulating during cooldown. The cooldown is usually fairly fast, so exterior heat gain is ignored. The calculation is based on the average refrigeration needed during the first half-cooling period. Water will often be considered cold enough for operation when it reaches seven-eighths cool.

$$R = [(V \cdot d \cdot c_w + m_c \cdot c_c) \cdot \Delta T)/(t \cdot k)] + Q_p/k$$

where

R = refrigeration capacity (kW or tons)

V = volume of water in reservoir (m^3 or ft^3)

d = density of water, 1,000 kg·m^{-3} or 62.4 lb of water per cubic foot

ΔT = water temperature drop during the first half-cooling period of cooling process (°C or °F)

c_w = heat capacity of water (4.187 kJ·kg^{-1}·°C^{-1} or 1.0 Btu/lb/°F)

m_c = weight of the empty cooler; includes all parts that are cooled (kg or lb)

c_c = specific heat of the cooler (assuming the cooler is made entirely of steel, its specific heat equals 0.46 kJ·kg^{-1}·°C^{-1} or 0.11 Btu/lb/°F)

t = half-cooling time (hours)

Q_p = heat (in Btuh) added by water pump (assume 90% of the motor energy use ends up as heat in the water and 1 hp = 2,540 Btuh; for example, a 30-hp pump motor adds 30 × 2,540 × 0.90 = 68,580 Btuh; a fully loaded 22-kW motor adds 19.8 kW of heat)

k = 3,600 s·h^{-1} or 12,000 Btuh per ton of refrigeration.

References

(ASHRAE) American Society of Heating, Refrigerating, and Air-Conditioning Engineers. 2022. ASHRAE handbook: Refrigeration. Atlanta: American Society of Heating, Refrigerating, and Air-Conditioning Engineers.

Baird, C. D., J. J. Gaffney, and M. T. Talbot. 1988. Design criteria for efficient and cost-effective forced air cooling systems for fruits and vegetables. ASHRAE Transactions 94(1):1434–1454.

Barger, W. R. 1963. Vacuum precooling: A comparison of the cooling of different vegetables. U.S. Department of Agriculture Market Quality Research Division Marketing Research Report 600. https://ia601901.us.archive.org/2/items/vacuumprecooling600barg/vacuumprecooling600barg.pdf

Becker, B. R., and Fricke, B. A. 1996. Transpiration and respiration of fruits and vegetables. In W. E. Murphy, ed., New developments in refrigeration for food safety and quality. Paris, France: International Institute of Refrigeration and St. Joseph, MI: American Society of Agricultural Engineers. 110–121.

Bennett, A. H. 1963. Thermal characteristics of peaches as related to hydrocooling. U.S. Department of Agriculture Technical Bulletin 1292.

Bergqvist, J., N. Dokoozlian, and N. Ebisuda. 2001. Sunlight exposure and temperature effects on berry growth and composition of cabernet sauvignon and grenache in the central San Joaquin Valley of California. American Journal of Enology and Viticulture 52(1):1–7. https://www.ajevonline.org/content/52/1/1

Buck, A. L. 1981. New equations for computing vapor pressure and enhancement factor. Journal of Applied Meteorology 20(12):1527–1532. https://doi.org/10.1175/1520-0450(1981)020<1527:NEFCVP>2.0.CO;2

Chau, K. V., R. A. Romero, C. D. Baird, and J. J. Gaffney. 1987. Transpiration coefficients of fruits and vegetables in refrigerated storage. Atlanta: American Society of Heating, Refrigerating, and Air-Conditioning Engineers report 370-RP.

Cheyney, C. C., R. F. Kasmire, and L. L. Morris. 1979. Vacuum cooling wrapped lettuce. California Agriculture 33(10):18–19. http://calag.ucanr.edu/Archive/?article=ca.v033n10p18

Crisosto, C., J. Smilanick, N. Dokoozlian, and D. Luvisi. 1994. Maintaining table grape post-harvest quality for long distance markets. Proceedings of the International Symposium on Table Grape Production.

De Castro, L. R., C. Vigneault, and L. A. B. Cortez. 2005. Cooling performance of horticultural produce in containers with peripheral openings. Postharvest Biology and Technology 38(3):254–261. https://doi.org/10.1016/j.postharvbio.2005.07.004

Dossat, R. J., and T. J. Horan. 2002. Principles of refrigeration. 5th ed. Upper Saddle River, NJ: Prentice Hall.

Ferrua, M. J., and R. P. Singh. 2009. Design guidelines for the forced-air cooling process of strawberries. International Journal of Refrigeration 32:1932–1943. https://doi.org/10.1016/j.ijrefrig.2009.07.011

Flockens, F. H., and H. F. T. Meffert. 1972. Biophysical properties of horticultural products as related to loss of moisture during cooling down. Journal of the Science of Food and Agriculture 23(3):285–298. https://doi.org/10.1002/jsfa.2740230304

Fricke, B. 2006. Precooling fruits and vegetables. ASHRAE Journal 48:20–28.

Gaffney, J. J., and C. D. Baird. 1975. Susceptibility of West Indian avocados to chilling injury as related to rapid cooling with low temperature air or water. Proceedings of the Florida State Horticultural Society 88:490–496. http://avocadosource.com/Journals/FSHSP/FSHSP_VOL_88_PG_490-496_1975.pdf

Gan, G., and J. L. Woods. 1989. A deep bed simulation of vegetable cooling. In V. A. Dodd and P. M. Grace, eds., Land and water use. Rotterdam: Balkema. 2301–2308.

Gentry, J. P., and K. E. Nelson. 1964. Conduction cooling of table grapes. American Journal of Enology and Viticulture 15(1):41–46. https://www.ajevonline.org/content/15/1/41

Guillou, R. 1958. Some engineering aspects of cooling fruits and vegetables. Transactions of the ASAE 1(1):38–39, 42.

———. 1960. Coolers for fruits and vegetables. Bulletin of the California Agricultural Experiment Station 773. 66.

———. 1963. Pressure cooling for fruits and vegetables. ASHRAE Journal 5(11):45–49.

Haerter, A. A. 1963. Flow distribution and pressure change along slotted or branched ducts. ASHRAE Journal 5(1):47–59.

Hardenburg, R. E., A. E. Watada, and C. Y. Yang. 1986. The commercial storage of fruits, vegetables, and florist and nursery stocks. U.S. Department of Agriculture Handbook 66.

Harvey, J. M. 1963. Improved techniques for vacuum cooling vegetables. ASHRAE Journal 5(1):41–44.

Hawkins, S. A. 2010. High efficiency apparatus and method for cooling produce. US patent 7,832,218.

Hruschka, H. W. 1977. Postharvest weight loss and shrivel in five fruits and five vegetables. U.S. Department of Agriculture Marketing Research Report 1059.

Kasmire, R. F., and R. A. Parsons. 1971. Precooling cantaloupes, a shipper's guide. Berkeley: Agricultural Extension Service.

Kitinoja, L., and J. F. Thompson. 2010. Pre-cooling systems for small-scale producers. Stewart Postharvest Review 2(2).

Luvisi, D. A., H. H. Shorey, J. F. Thompson, T. Hinsch, and D. C. Slaughter. 1995. Packaging California table grapes. Oakland: UC Agriculture and Natural Resources Publication 1934. https://anrcatalog.ucanr.edu/pdf/1934.pdf

Mitchell, F. G., R. Guillou, and R. A. Parsons. 1972. Commercial cooling of fruits and vegetables. UC Agricultural Experiment Station Manual 43.

Mitchell, F. G., R. A. Parsons, and G. Mayer. 1971. Cooling trials with plastic tray pack nectarines in various containers. California Agriculture 25(9):13–15. http://calag.ucanr.edu/Archive/?article=ca.v025n09p13

Mitchell, F. G., N. F. Sommer, J. P. Gentry, R. Guillou, and G. Mayer. 1968. Tight-fill fruit packing. California Agricultural Experiment Station Circular 548. https://archive.org/details/tightfillfruitpa548mitc/mode/2up

Nelson, K. E. 1985. Harvesting and handling California table grapes for market. Oakland: UC Agriculture and Natural Resources Publication 1913. https://anrcatalog.ucanr.edu/Details.aspx?itemNo=1913

Ngcobo, M. E. K., M. A. Delele, U. L. Opara, C. J. Zietsman, and C. J. Meyer. 2012. Resistance to airflow and cooling patterns through multi-scale packaging of table grapes. International Journal of Refrigeration 35(2):445–452. https://doi.org/10.1016/j.ijrefrig.2011.11.008

O'Sullivan, J., M. J. Ferrua, R. Love, P. Verboven, B. Nicolaï, and A. East. 2016. Modelling the forced-air cooling mechanisms and performance of polylined horticultural produce. Postharvest Biology and Technology 120:67–75. https://doi.org/10.1016/j.postharvbio.2016.05.008

Pacific Gas and Electric. 2008. Field vacuum precooling evaluation. https://www.etcc-ca.com/reports/field-vacuum-precooling

Parsons, R. A., F. G. Mitchell, and G. Mayer. 1972a. Forced-air cooling of fruit in bulk bins. American Society of Agricultural Engineers Special Publication SP-01-72:38–41.

———. 1972b. Forced-air cooling of palletized fresh fruit. Transactions of the ASAE 15:729–731.

Pentzer, W. T., R. L. Perry, G. C. Hanna, J. S. Wiant, and C. E. Asbury. 1936. Precooling and shipping California asparagus. UC Agricultural Experiment Station Bulletin 600.

Perry, R. L., and R. M. Perkins. 1968. Hydrocooling sweet corn. St. Joseph, MI: American Society of Agricultural Engineers Paper 68–800.

Rennie, T. J., V. G. S. Raghavan, C. Vigneault, and Y. Gariépy. 2001. Vacuum cooling of lettuce with various rates of pressure reduction. Transactions of the ASAE 44(1):89–93. https://elibrary.asabe.org/abstract.asp?aid=2292

Rij, R. E., J. F. Thompson, and D. S. Farnham. 1979. Handling, precooling, and temperature management of cut flower crops for truck transportation. U.S. Department of Agriculture Science and Education Administration AAT-W-5. https://ucanr.edu/datastoreFiles/234-1052.PDF

Ritenour, M. A., and C. H. Crisosto. 1996. Hydrocooler water sanitation in the San Joaquin Valley stone fruit industry. Central Valley Postharvest Newsletter 5(1):15–19.

Robinson, J. E., K. M. Browne, and W. G. Burton. 1975. Storage characteristics of some vegetables and soft fruits. Annals of Applied Biology 81(3):399–408. https://doi.org/10.1111/j.1744-7348.1975.tb01656.x

Sainsbury, G. F. 1961. Cooling apples and pears in storage rooms. U.S. Department of Agriculture Marketing Research Report 474. https://doi.org/10.22004/ag.econ.312604

Sastry, S. K., C. D. Baird, and D. E. Buffington. 1978. Transpiration rates of certain fruits and vegetables. ASHRAE Transactions 84(1):237–255.

Stewart, K. S., and H. M. Couey. 1963. Hydrocooling vegetables: A practical guide to predicting final temperatures and cooling times. U.S. Department of Agriculture Marketing Service Marketing Research Report 637.

Stewart, J. K., and W. J. Lipton. 1960. Factors influencing heat loss in cantaloups during hydrocooling. U.S. Department of Agriculture Marketing Research Report 421.

Stoecker, W. F. 1988–1995. Industrial refrigeration. Vols. 1–2. Troy, NY: Business News Publishing.

Tanner, D. J., A. C. Cleland, and T. R. Robertson. 2002. A generalised mathematical modelling methodology for design of horticultural food packages exposed to refrigerated conditions: Part 3, mass transfer modelling and testing. International Journal of Refrigeration 25(1):54–65. https://doi.org/10.1016/S0140-7007(01)00017-2

Teruel, B., L. A. Cortez, P. Leal, and A. G. B. Lima. 2001. Estudo teórico do resfriamento com ar forçado de frutas de geometrias diferentes. Revista Ciência e Tecnologia dos Alimentos 21(2):228–235. http://www.scielo.br/pdf/cta/v21n2/7472.pdf

Wade, N. L. 1984. Estimation of the refrigeration capacity required to cool horticultural produce. International Journal of Refrigeration 7:358–366.

Chapter 2
Refrigerated storage

James F. Thompson

Orderly marketing of perishable commodities often requires refrigerated storage to balance fluctuations between rates of produce harvest and sales. This period of storage may last from a few days to many months. Long-term storage can extend marketing beyond the end of the harvest season. The goals of storage are to

- slow the produce biological activity by maintaining the lowest temperature that will not cause freezing or chilling injury—and by controlling atmospheric composition of oxygen, carbon dioxide, and ethylene
- slow the growth and spread of microorganisms by maintaining low temperatures and minimizing condensation of moisture on produce
- reduce produce moisture loss and the wilting and shriveling that result from it by reducing the temperature difference between the produce and air in the storage room air and by maintaining high humidity in the storage room

The storage facility may also be used to apply special treatments to some commodities. For example, potatoes and sweet potatoes are held for a few days at high temperature and high relative humidity to cure wounds sustained during harvest. Table grapes are fumigated with sulfur dioxide to minimize *Botrytis* decay damage. Pears may be held near room temperature and exposed to ethylene for 24 hours to ripen more quickly and uniformly. This chapter describes the equipment and techniques commonly used in storage facilities to control temperature, relative humidity, and atmospheric composition (fig. 2.1).

Temperature

The storage room should be kept within ±1°C (2°F) of the desired temperature for the commodities being stored. A narrower range may be needed for storage temperatures very close to the freezing point of the commodity. Temperatures below the optimal range can cause freezing or chilling injury, while temperatures above it can shorten storage life. In addition, wide temperature fluctuations can result in condensation

Figure 2.1. Interior view of a table-grape cold storage.

of water on, and faster water loss from, commodities. Recommended temperatures and relative humidity for long-term storage of horticultural produce are listed in appendix A. In many storage facilities, particularly at wholesale distribution centers and retail stores, many different produce items are held in a common room. Table 2.1 groups commodities into four temperature and relative humidity groups for short-term storage.

Maintaining temperatures within the prescribed range depends on several important design factors. The refrigeration system must be able to handle the maximum expected heat load. Undersized systems allow air temperature to rise during peak heat-load conditions, while an oversized system is unnecessarily expensive. The system should also be designed so that air leaving the refrigeration coils is close to the desired storage temperature. This prevents large air

Table 2.1. Compatible fresh fruit and vegetables during 10-day storage

Group 1A: 0–2°C (32–36°F) and 90–98% relative humidity				
Vegetables				
alfalfa sprouts	broccoflower*	cut vegetables	kohlrabi	salsify
amaranth	Brussels sprouts	daikon*	leeks*	scorzonera
anise	cabbage*	endive* (chicory)	lettuce*	shallots
artichokes	carrots*	escarole*	mint	snow peas*
arugula*	cauliflower*	fennel	mushrooms	spinach*
asparagus*	celeriac	garlic	mustard greens*	sweet peas*
beans: fava, lima	celery*	green onions*	parsley*	Swiss chard
bean sprouts	chard*	herbs* (not basil)	parsnips	turnips
beets	Chinese cabbage	horseradish	radicchio	turnip greens*
Belgian endive*	Chinese turnips	Jerusalem artichokes	radishes	water chestnuts
bok choy	collards*	kailon	rutabagas	watercress*
broccoli*	corn: sweet, baby	kale	rhubarb	
Group 1B: 0–2°C (32–36°F) and 85–95% relative humidity				
Fruit and melons				
apples	cantaloupes	elderberries	lychees	pomegranates
apricots	cashew apples	figs	nectarines	prunes
avocados, ripe	cherries	gooseberries	peaches	quinces
Barbados cherries	coconuts	grapes	pears, Asian	raspberries
blackberries	currants	kiwifruit*	pears, European	strawberries
blueberries	cut fruit	loganberries	persimmons*	—
boysenberries	dates	longans	plums	—
caimitos	dewberries	loquats	plumcots	—

Continued on next page

Table 2.1. Compatible fresh fruit and vegetables during 10-day storage, continued

Group 2: 7–10°C (45–50°F) and 85–95% relative humidity				
Vegetables		**Fruit and melons**		
basil	kiwanos (horned melons)	avocados, unripe	guavas	pepinos
beans, green	long beans	babacos	Juan Canary melons	pineapples
beans, snap	malangas	cactus pears, tuna	kumquats	pomelos
beans, wax	okra*	calamondins	lemons*	sugar apples
cactus leaves (nopales)	peppers: bell, chili	carambolas	limes*	tamarillos
calabazas	squashes, soft rind*	cranberries	limequats	tamarind
chayote*	squashes, summer	custard apples	mandarins	tangelos
cowpeas (Southern pea)	tomatillos	durians	olives	tangerines
cucumbers*	winged beans	feijoas	oranges	ugli fruit
eggplant*	—	granadillas	passion fruit	watermelons
—	—	grapefruit*	—	—
Group 3: 13–18°C (55–65°F) and 85–95% relative humidity				
Vegetables		**Fruit and melons**		
bitter melons	pumpkins	atemoyas	honeydew melons	Persian melons
boniatos	squashes, hard rind*	bananas	jaboticabas	plantains
cassava	squashes, winter*	breadfruit	jackfruit	rambutans
dry onions	sweet potatoes*	canistels	mamey sapotes	sapodillas
ginger	taro (dasheen)	casaba melons	mangoes	sapotes
jicamas	tomatoes, all maturities	cherimoyas	mangosteens	soursops
potatoes	yams*	crenshaw melons	papayas	—

Note: "—" indicates that table cells are left intentionally blank.

Note: Ethylene should be kept below 1 ppm in storage areas.

*Products sensitive to ethylene

temperature fluctuations as the refrigeration system cycles on and off. Large refrigeration coils installed with equipment to control the refrigerant pressure in the coils (suction pressure control) allow a small temperature difference between the air leaving the coils and the air in the room, while still offering adequate refrigeration capacity. A small temperature difference also increases the relative humidity in the room and may reduce frost build-up on the coils. A large free space above the stored produce allows air from the evaporators to mix with room air before coming in contact with the produce.

Temperature variation is minimized by having adequate air circulation. Most storages are designed to provide an airflow of 0.052 cubic meters per second and per metric ton (100 cfm/ton) of produce, based on the maximum amount of produce that can be stored in the room. This level of airflow is needed to cool the produce to the desired storage temperature, and may be needed if the produce has a high respiration rate (that is, a high rate of heat production). Since high rates of airflow can cause excessive produce weight loss, and because fans are a significant source of heat, the system should be designed to reduce airflow to 0.0208 to 0.0104 cubic meters per second and per metric ton (20–40 cfm/ton) after the produce has reached storage temperature. Systems that control motor speed, such as variable frequency controllers for AC motors, are often used to control fan speed. Operating fans at the lowest possible speed minimizes addition of heat to the storage room. The warmest produce will tend to be near the top of the room next to a warm wall or roof, and farthest from the evaporator fans. Operating fans at reduced speeds has the disadvantage of reducing the distance air travels after it leaves the fan. In some cases, that distance may not be enough to distribute cold air to the farthest sections of the storage, resulting in increased temperature variability in the cold room—which can include the possibility of freezing damage in some areas of the cold room. An alternative approach is to operate the fans at normal speed but cycle them off periodically. This is called fan cycling and is often used in controlled-atmosphere storages.

Low circulation rates require that the system be designed to move air uniformly past all the stored produce. Figure 2.2 shows some systems used to uniformly distribute air in large storage rooms. When airflow is directed around the sides of bins or pallet loads, they must be stacked to form air channels 10 to 15 centimeters (4–6 in) wide around the sides of each unit. These channels should be formed parallel to the direction of air movement. There should also

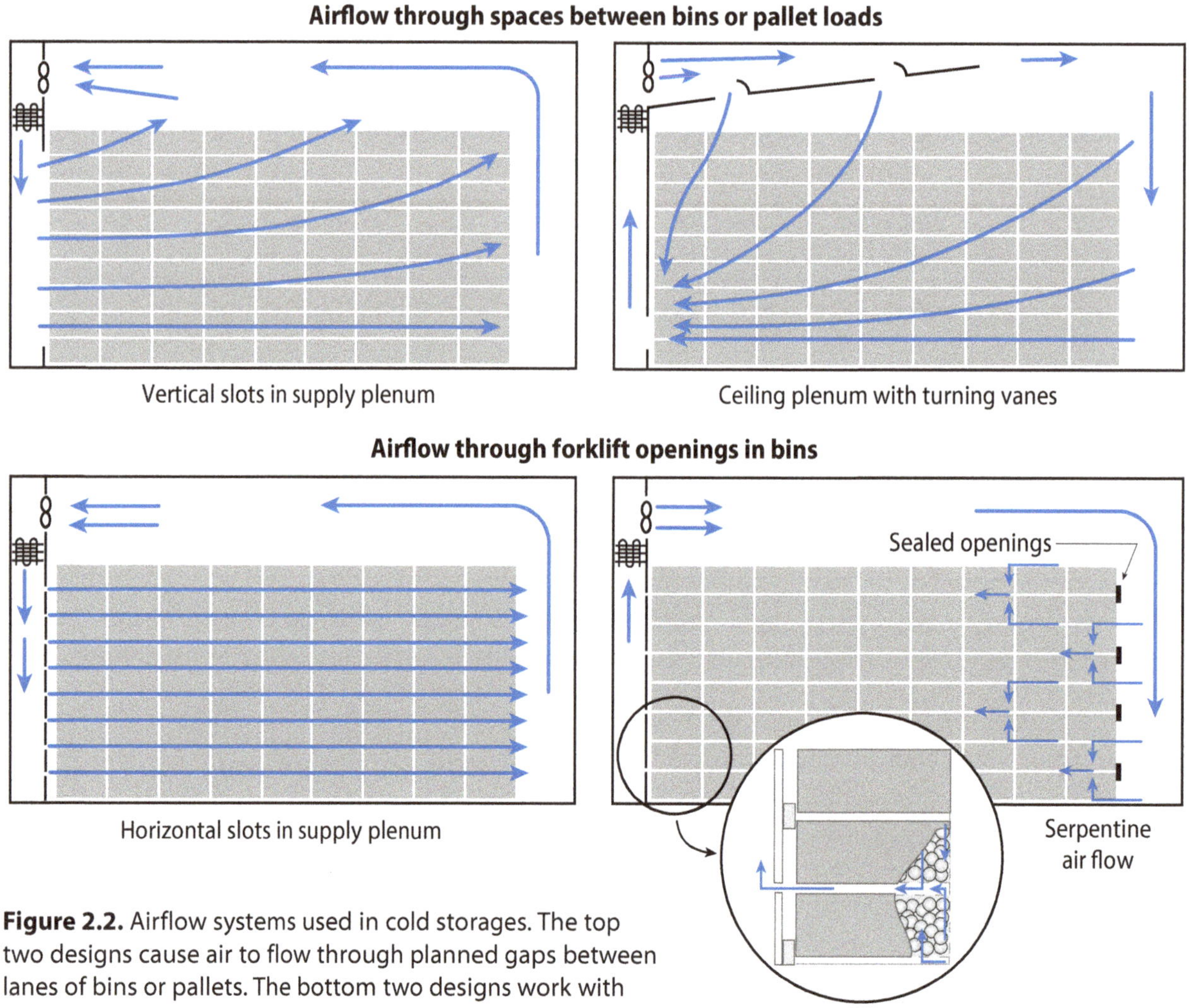

Figure 2.2. Airflow systems used in cold storages. The top two designs cause air to flow through planned gaps between lanes of bins or pallets. The bottom two designs work with tightly stacked bins and force air through forklift openings.

be space between the pallets or bins and the walls to allow refrigerated air to absorb the heat conducted from the outside through the walls. Curbs installed on the floor next to walls help insure there is an air space between pallets and the walls. When forklift openings are used for air distribution, they must align along the entire length of the air path. The serpentine airflow system is the only one that forces air to flow through the produce. This speeds initial cooling but may not be necessary for uniform produce temperatures in long-term storage.

Because air takes the path of least resistance, partially filled rooms often have poor air distribution. Large rooms can be divided into sections, sometimes called bays, with uninsulated walls running parallel to the direction of airflow. This allows produce in one bay to be removed with little effect on airflow in neighboring bays.

Smaller rooms with packaged fan and evaporator coils use an airflow pattern similar to the ceiling plenum design. The evaporator unit is installed near the ceiling and high-capacity evaporator fans can discharge air up to 15 meters (50 ft) along the underside of the ceiling. Air returning to the evaporator flows horizontally back past the produce and then up to the evaporator in a plenum formed by stacking produce 20 to 25 centimeters (8–10 in) away from the wall under the evaporator. Rooms cooled with roof-mounted packaged evaporative coolers can use exhaust vents near the floor to distribute air downward past produce bins.

In some long-term controlled-atmosphere storages, airflow is minimized by cycling evaporator fans off for as much as 85 to 90 percent of the time. In the winter in temperate climates, most refrigeration demand comes from heat released by evaporator fans. Reducing fan operation time reduces heat input, which in turn reduces refrigeration operation. Tests conducted in controlled atmosphere storages in the northwestern United States showed that fan cycling reduced electricity use by 50 to 55 percent (Koca and Hellickson 1993). It may also increase humidity in the storage and reduce produce moisture loss.

Thermostat sensors are usually placed 1.5 meters (5 ft) above the floor, for ease of checking the sensors, and in representative locations around the room. Thermostat sensors should not be placed near sources of heat, such as doors or walls with an exterior surface. Nor should they be placed in a cold area, such as a location near the air discharge of the refrigeration unit. A calibrated thermometer should be used to periodically check the thermostat. Air temperatures should be regularly logged or recorded automatically. Errors of only a few degrees can affect produce quality.

Humidity

For most perishable commodities, the relative humidity in a long-term storage facility should be kept at 90 to 95 percent. Relative humidity below this range results in unacceptable moisture loss, while relative humidity very close to 100 percent may cause excessive growth of microorganisms and surface cracking on some fruit. It is unusual for the relative humidity in a storage facility to be too high—though some controlled-atmosphere storages have relative humidities slightly above 95 percent, and in those cases fruit must be warmed before packing because it is so turgid.

Fiberboard containers are weakened by prolonged exposure to the high relative humidity recommended for most horticultural commodities. Typical fiberboard exposed to 95 percent relative humidity has half as much strength as the same material exposed to 50 percent relative humidity. The storage humidity for perishables can be lowered below recommended levels if they are packaged in plastic bags or box liners to retard moisture loss.

Refrigeration equipment must be specially designed to maintain high relative humidity. In systems not designed for horticultural commodities, the evaporator coils (which produce the cold air) operate at a temperature about 6°C (11°F) lower than the desired air temperature in the room. This causes an excessive amount of moisture to condense on the coils and can result in 70 to 80 percent relative humidity in the storage room. Cold air leaving the evaporator coils is at 100 percent relative humidity, but it warms as it removes heat from the storage. Since warm air can hold more water vapor than cold air, its relative humidity decreases. Evaporator coils should be selected to produce the highest possible air temperature while still remaining capable of keeping the cold room at its set-point air temperature. This is done by selecting coils with a large surface area and not allowing them to operate more than 3°C (5°F) colder than the room air temperature. The control system should also be designed to operate the coils at their highest possible temperature at all times. This is often done with equipment that controls the refrigerant pressure in the evaporator (suction pressure control).

Mechanical humidifiers or fog spray nozzles are sometimes used to add moisture to the storage room and reduce the drying effect of the evaporator coils. However, this added moisture results in the need for more frequent defrosting of evaporator coils. Humidifiers are needed when produce is stored at a temperature warmer than the outside environment, such as in cold-winter locations where heat must be added to prevent damage by low temperatures.

Some refrigeration systems use a wetted coil heat exchanger to maintain humidity. In this system, water is cooled to 0°C (32°F), or a higher temperature if higher room temperatures are desired. The water is sprayed down through a coil, and the storage area air is cooled and humidified to nearly 100 percent relative humidity as it moves upward through the wetted coil. However, the air picks up heat as it moves through the storage area, and the rise in temperature reduces relative humidity. This system is usually limited to situations in which air temperatures are kept above 0.2°C (32.4°F). It does not work well for commodities that are held very close to or below 0°C—unless compounds that lower the freezing point of water, such as caustic soda, are used in the refrigerated water.

Partially dried produce such as garlic, ginger, and dry onion are held at 65 to 75 percent relative humidity. Dried fruit and nuts should be held at 55 to 65 percent relative humidity. These low humidities are obtained by installing a small evaporator coil, designed to operate with a large temperature difference between the refrigerant and the temperature of the air leaving the coil. Sometimes the refrigeration coil is installed with a reheat system. In the first part of the coil, the air is cooled considerably below the desired room temperature to dry the air to the desired dew-point temperature. Then, in a reheat section, the air is heated to the set-point air temperature. Heat is supplied by an electrical resistance heater—or by hot refrigerant gas from the compressor discharge, piped to a reheat section of the evaporator coil.

Refrigeration

To maintain a desired storage temperature, the refrigeration capacity of a system must be designed to handle heat inputs to a storage volume (including heat conducted through walls, floor, and ceiling); field and respiration heat from the produce; heat from air infiltration; and heat from personnel and equipment such as lights, fans, and forklifts. Details of heat-load calculations are described in ASHRAE (2021) and in Stoecker (1998).

Refrigeration equipment for storage facilities is generally not designed to remove much field heat from the produce because doing so would require a large refrigeration capacity; a separate initial cooling process is used for this purpose.

Mechanical refrigeration

Most storage facilities use mechanical refrigeration to maintain the temperature in a cold room. Mechanical refrigeration systems are based on the fact that refrigerants absorb heat as they undergo a phase transition from liquid to gas. The simplest cooling method involves a controlled release of liquid nitrogen or liquid carbon dioxide. As these liquids evaporate, they absorb heat from the storage area. However, these methods require a constant outside supply of refrigerant and are only used to a limited extent, with highway vans and railcars. The more common mechanical refrigeration systems use a refrigerant such as ammonia or a variety of halocarbon fluids, whose vapor can be easily recaptured in liquid form by a compressor and condenser system.

Figure 2.3 shows the components of a typical vapor-recompression (or mechanical) refrigeration system. The refrigerant fluid passes through the expansion valve, where the pressure drops, causing the liquid to evaporate at temperatures low enough to be effective in removing heat from the storage area. Heat from the material to be cooled is transferred to the room air, which is then forced past the evaporator (that is, the cooling coil located in the room), usually a finned tube heat exchanger—which transfers the heat from the air to the refrigerant, causing it to evaporate. After fully changing to a gas, the refrigerant is repressurized by the compressor and then passed through a condenser, where it is cooled to a liquid. The condenser is located outside the storage area and releases heat. The condensed liquid is stored in the receiver and is metered out as needed for cooling.

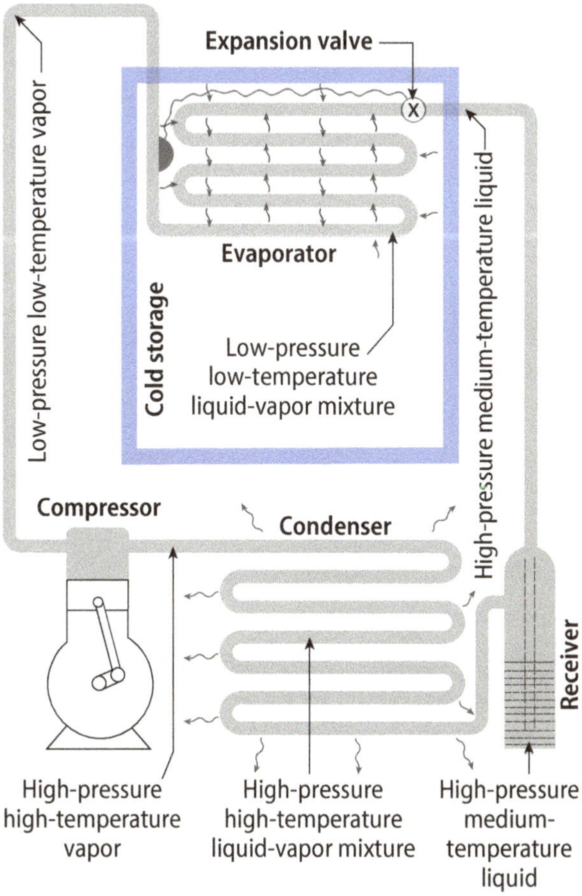

Figure 2.3. Schematic of a typical vapor-recompression or mechanical refrigeration system.

Expansion valves

Small mechanical refrigeration systems are controlled primarily by expansion valves, which regulate the pressure of the refrigerant in the evaporator. Low pressures cause the liquid refrigerant to evaporate at low temperatures. The valve also controls the flow of refrigerant, which affects the amount of refrigeration capacity available. Capillary tubes and thermostatic expansion valves are the two most common types of expansion valves.

The capillary tube is used with very small refrigeration equipment (less than 1 horsepower). It is a tube 0.6 to 6 meters (2–20 ft) long with a very small inside diameter of 0.6 to 2.3 millimeters (0.024–0.091 in). The resistance of the liquid flowing through the tube creates the needed pressure drop between the low-pressure and high-pressure sides of the system and regulates the flow of refrigerant. A capillary is inexpensive and has no moving parts to maintain, but it cannot be adjusted, is subject to clogging, and requires a relatively constant weight of refrigerant in the system.

A thermostatic expansion valve regulates the flow of refrigerant to maintain a constant temperature difference between the evaporator inlet (or evaporating temperature) and the coil outlet, maintaining a constant refrigerant superheat. It is commonly used in smaller, commercial-scale refrigeration systems. It allows the low-side pressure to vary, so that when high refrigeration loads are required, the temperature of the evaporator coil increases. This type of expansion valve is not well suited to obtaining the high relative humidity needed in long-term produce storage.

Large refrigeration systems may use a flooded coil, an evaporator coil that is designed to always have liquid refrigerant in it. A flooded coil has greater heat transfer efficiency than a nonflooded coil of equal size. Refrigerant flow is controlled with a surge tank installed before the coil, which maintains a constant liquid refrigerant level in the coil.

A thermostatic expansion valve may operate in parallel with the float valve to control refrigerant pressure in the evaporator. Other controls, such as suction pressure regulators, may also be used in conjunction with float controls. These are especially useful in maintaining the highest possible evaporator coil temperature in order to maintain high relative humidity in the storage room.

Evaporators

The liquid refrigerant absorbs heat and vaporizes in the evaporator. Modern cold storages use finned tube evaporators (fig. 2.4). Air from the storage is forced past the tubes by fans, which are part of a complete evaporator unit. Evaporators operating below 0°C (32°F) build up frost that must be periodically removed to maintain good heat transfer efficiency. Evaporators may be defrosted by flooding the coils with water, by using electric heaters, by directing hot refrigerant gas to the evaporators, or by continuously defrosting with a brine or glycol solution flowing over the coils.

Maintaining the highest possible refrigerant pressure (suction pressure) in the evaporator coils reduces compressor energy use. This can be done by using large evaporator coils and a control system that maintains the highest possible suction pressure.

Compressors

The first step in reusing the refrigerant is using a compressor to compress the cool gaseous refrigerant to a hot gas. The most common types of refrigeration compressors are reciprocating (piston) and rotary screw (fig. 2.5). Reciprocating compressors come in a wide range of sizes and can be set up to operate efficiently at varying refrigerant flow rates. Flow rates are varied by shutting off pairs of cylinders in a unit, which may have six to twelve cylinders. The main disadvantage of reciprocating compressors is their fairly high maintenance costs. Rotary screw compressors have low maintenance costs but are not available in sizes smaller than about 23 kilowatts (31 hp).

Using a compressor system that operates efficiently over the required range of refrigerant flows is also recommended. Screw compressors operate efficiently only at flow rates near their maximum capacity. In large operations, several screw compressors are installed in parallel, allowing some to be shut down as refrigeration demand drops. Some vacuum coolers are operated with a refrigeration system that also operates cold rooms. An optimum design uses a subsystem to provide the −14° to −11°C (6–12°F) refrigerant needed for the vacuum cooler, while

Figure 2.4. Finned tube evaporators. *Photo:* James F. Thompson.

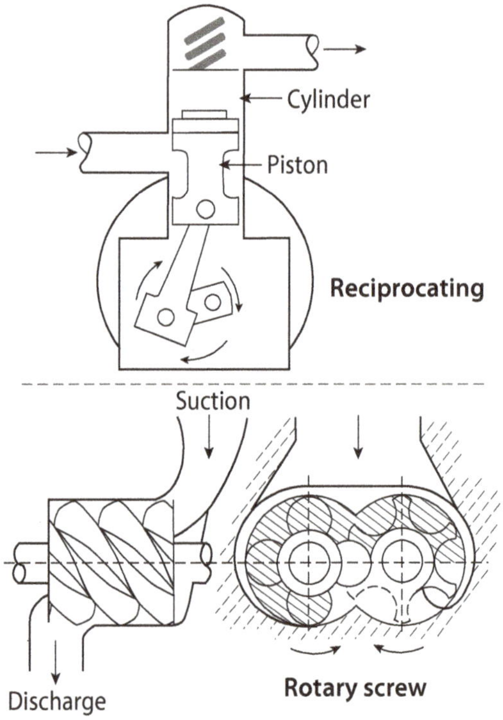

Figure 2.5. Common types of refrigeration compressors.

the main refrigeration system provides higher-pressure refrigerant for the cold rooms.

Condensers

The hot refrigerant vapor loses heat and condenses into a liquid in the condenser (see fig. 2.3). Condensers are categorized as air-cooled or water-cooled. Small systems usually use an air-cooled unit. Many home refrigerators, for instance, have a fan that blows air past a finned tube bundle under the unit. Large condensers are more likely to be water-cooled. Water is a better heat conductor than air, allowing water-cooled condensers to be smaller than forced-air units of equal capacity. However, water-cooled units may require significant quantities of water, which can be expensive to obtain and dispose of. Evaporative condensers use much less water than do water-cooled condensers. They require close attention to water quality to maintain efficiency and to prevent damage to the heat exchanger. In 2018, condenser manufacturers began offering adiabatic condensers. These units first send outside air through a water-wetted packed bed that evaporatively cools the air, after which it flows through the condensing coils. The condenser is larger than a traditional evaporative condenser of the same capacity but uses less water.

Energy use is minimized by selecting a condenser that cools the refrigerant fluid to as low a temperature as possible. For example, a facility maintaining air temperature of 0°C (32°F) and a condensing temperature of 52°C (125°F) requires 50 percent more power than one that operates at a condensing temperature of 35°C (95°F). In warm areas, well water–cooled, evaporatively cooled, or adiabatic condensers should be selected over air-cooled units.

Refrigerants

The choice of which refrigerant to use in a mechanical refrigeration system is based on the following factors:

- cost of refrigerant (halocarbon refrigerants are more expensive than ammonia)
- environmental regulations (ozone depletion and the global warming effects of many halocarbon refrigerants restrict their availability)
- compatibility (ammonia cannot be used with metals that contain copper; halocarbon refrigerants cannot be used with alloys containing more than 2% magnesium and may damage some elastomeric materials)
- toxicity (ammonia at very low concentrations can injure perishable commodities; ammonia is toxic to humans)
- flammability (ammonia is flammable at 16–28% mixtures by volume in air and with an ignition source greater than 540°C (1,000°F); most commonly used halocarbon refrigerants are not flammable, or have a low flammability)

Control systems

A large refrigeration system requires a good control system and equipment for displaying the refrigeration system's operating conditions. At a minimum, lights or gauges should be installed to indicate the operating status of fans and compressors and the fluid levels in surge and receiver tanks. Controls should be set up to allow manual operation of motors.

Microcomputers and programmable controllers allow even more precise control of large refrigeration systems. They are especially valuable in reducing electricity use during peak-rate periods. Defrost cycles can be programmed to occur at night, and unnecessary fans and compressor motors can be turned off during peak-rate periods.

Absorption refrigeration

Absorption refrigeration, used in a few cold-storage operations, differs from mechanical refrigeration in that the vapor is recovered primarily through use of heat rather than mechanical power. It is less energy-efficient than mechanical refrigeration and usually is used only where an inexpensive source of heat is available. Processing facilities with excess low-pressure steam are well suited to use absorption refrigeration.

Secondary refrigerant

Some storage rooms are cooled with a secondary coolant. A brine solution (sodium chloride or calcium chloride) or glycol (propylene glycol or ethylene glycol) is cooled by a mechanical refrigeration system and pumped to heat exchangers in the cold storage. These systems are a little less energy-efficient than conventional systems. However, they dramatically reduce the quantity of first-stage refrigerant needed and confine it to the engine room. This is a great asset in dealing with the flammability and safety issues of ammonia. Secondary refrigerant piping does not need to withstand the pressure of primary refrigerants, and plastic piping can sometimes be used. The temperature of the heat exchanger can be precisely controlled with a mixing valve. Brines and glycol coolant solutions are corrosive and must be used with corrosion inhibitors, and they should never come in contact with zinc. Food-grade sodium chloride and propylene glycol should be used for any facility handling food for human consumption.

Alternative refrigeration sources

In many developing countries, where mechanical refrigeration is prohibitively expensive to install, maintain, and operate, a number of other refrigeration techniques can be used. In some cases, these techniques can provide cooling levels that approach recommended storage conditions. In others, they are a compromise between proper storage conditions and costs of equipment, capital, and operations.

Room air conditioners

Normal wall- or window-mounted air conditioners cannot produce air temperatures low enough for most produce storage. They are also designed to produce air with low relative humidity, for human comfort. An easy-to-install controller (see, for example, storeitcold.com) allows air conditioners to cool air below 2°C (36°F) and also limits the amount of water vapor removed from the air. The cost of an air conditioner and controller is about 90 percent less than the purchase and installation cost of an equivalently sized commercial refrigeration system (Kitinoja and Thompson 2010). The controller will work with many but not all models of air conditioners. Check with the controller supplier before purchasing an air conditioner.

Air conditioners have refrigeration capacities in the range of 2.9 to 7.3 kilowatts (10,000–25,000 British thermal units per hour, or Btuh) and have the capacity to refrigerate well-insulated rooms from 4.6 to 13 square meters (50–140 ft^2). It is possible to install several air conditioners in larger rooms.

Evaporative cooling

Evaporative cooling techniques are very energy-efficient and economical (fig. 2.6). A well-designed evaporative cooler produces air with a

Figure 2.6. Evaporatively cooled sweet potato storage in California. Unit evaporators on the roof supply cool air to the storage. Room air is exhausted through vents along the bottom of the sidewalls.

relative humidity greater than 90 percent. The main limitation of evaporative coolers is that they cool air only to a few degrees above the wet-bulb temperature of the outside air. During the harvest season in the United States, wet-bulb temperatures vary from 10° to 25°C (50–77°F) depending on location, time of day, and weather conditions. This temperature range is acceptable for the storage of some chilling-sensitive commodities.

The water for evaporative cooling systems comes from domestic sources. It is also possible to cool by evaporating water directly from the commodity. Southern peas have been cooled in transit by erecting an air scoop above the cab of the truck that forces outside air through a bulk load of beans (Prussia and Shewfelt 1985). This system prevents heat build-up and keeps the beans at or below the outside air temperature. In order to avoid excessive water loss from the produce, it is not advisable to use this system for any great length of time.

Nighttime cooling

In some parts of the world, significant differences between night and day temperatures allow nighttime ventilation to be a means of refrigeration. In dry Mediterranean or desert climates, the difference between daily maximum and minimum temperatures can be as great as 22°C (40°F). Nighttime cooling is commonly used for unrefrigerated storage of potatoes, onions, sweet potatoes, hard-rind squashes, and pumpkins. As a rule, night ventilation effectively maintains a given produce temperature when the outside air temperature is below the produce temperature for 5 to 7 hours per day.

Produce temperature in the field is usually near air temperature at night. Harvesting during early morning hours, before sunrise, can be used to reduce field heat. Some growers in California use artificial lighting to allow night harvest.

It is theoretically possible to produce air temperatures below nighttime minimums by radiating heat to a clear nighttime sky. A clear night sky is very cold, and a good radiating surface such as a black metal roof can cool below the air temperature. In one study, unglazed solar collectors operated at night with clear skies produced water several degrees Celsius below ambient air temperature (Anderson et al. 2013). This concept is rarely used.

Naturally formed ice

Before the development of mechanical refrigeration, refrigeration was provided by ice harvested from shallow ponds or lakes during the winter. The ice was stored in straw and hauled to cities as needed during spring and summer. Energy costs today make it infeasible to transport ice any significant distance. However, cooling facilities in areas with cold winters can store ice nearby for summer use.

High-altitude cooling

High altitude can also be a source of cold. As a rule of thumb, air temperatures decrease by 6.4°C with every 1 kilometer (3.5°F per 1,000 ft) of increase in altitude. It is not possible to bring this air down to ground level because the air heats naturally due to compression as it drops in altitude. However, in some cases it may be possible to store commodities at high altitudes in mountainous areas. In California, for example, most perishable commodities are grown in the valley floors near sea level. However, much of the production is shipped east across the Sierra Nevada over passes about 1,800 meters (6,000 ft) high. Air temperature has the potential to be 10.2°C (21°F) cooler at these elevations, and it may reduce energy costs to store perishables there rather than on the valley floor.

Underground storage

Cellars, abandoned mines, and other underground spaces have been used for centuries to store fruit and vegetables. The temperature of the ground deeper than about 2 meters (6 ft) below the surface is equal to the average annual air temperature. Underground spaces work well for storing produce that has already been cooled, but not for removing field heat. Soil is a poor conductor of heat; once an underground storage area warms up, it does not cool down rapidly.

Well water

In some areas, well water can be an effective source of refrigeration. Well water temperature is often very near the average annual air temperature.

The storage building

The storage building must be sized to handle peak amounts of produce. The floor area can be calculated by dividing the volume of the produce by the maximum produce storage height and adding floor area for aisles, forklift maneuvering, and staging areas. The maximum storage height can be increased by using shelves or racks and forklifts with suitable masts. Multistory structures are generally not used for storage because of the difficulty and expense of moving the produce between levels.

The storage building should ideally have a floor perimeter in the shape of a square. A rectangular configuration has more wall area per square foot of floor area, resulting in higher construction costs and higher heat gain than a square configuration. Entrances, exits, and storage areas should be arranged so that produce generally moves in one direction through the facility, especially if the storage facility is used in conjunction with an initial cooling system to remove field heat.

Site selection

Good utility service must be available for the facility. Extending roads and energy utilities to a facility can be very expensive. Three-phase electrical power is needed to operate refrigeration equipment motors. In some areas a backup power supply may be advisable. There should be enough water to supply the evaporative condensers, personnel needs, and the needs of a packinghouse, if it is a part of the project. Consider the availability of fire protection services, gas supply, and sewer utilities. The area should have good drainage and room for future expansion. There should be enough space around the facility for movement of large highway trucks.

Building layout

Room layouts with an interior corridor offer better operating conditions for cold storages and better control of controlled-atmosphere storages than designs allowing room access only through exterior doors. The interior corridor, seen in layouts 1, 2, and 3 in figure 2.7, allows easy access to piping and controls. Doors and equipment are shielded from the elements and produce observation is easier. Layout 1 is common in small operations where storage and packing are done in one building. Layout 2 allows better produce flow than layout 3, but layout 3 has less area devoted to corridor. Layout 4 is the least expensive of the designs because none of the cold-storage building is used as a permanent corridor.

Refrigerated facilities can be constructed from a wide variety of materials. Floors and

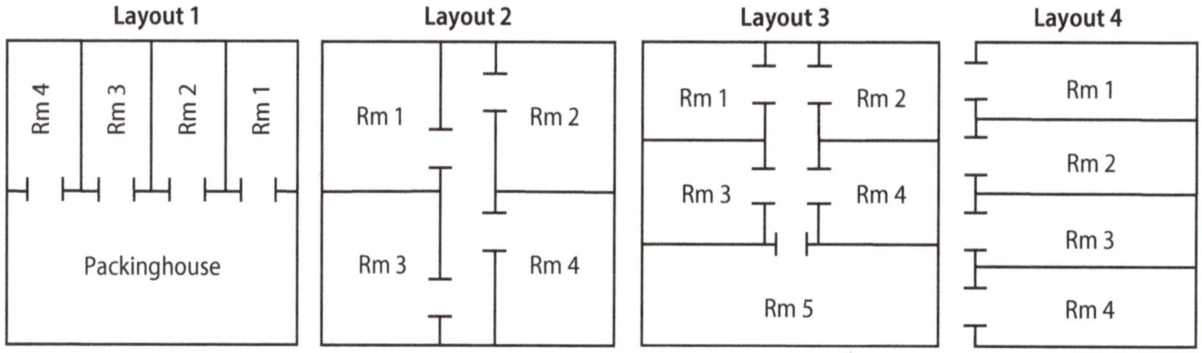

Figure 2.7. Typical layouts of cold-storage facilities.

foundations are usually concrete. A vapor barrier is installed to prevent moisture from moving through the floor and into the refrigerated space. Rigid insulation is sometimes placed above the barrier and below the concrete. A less expensive option for reducing heat loss through the floor is to install board insulation vertically along the outside of the perimeter foundation. Walls can be made of concrete blocks, tilt-up concrete panels, insulated metal panels, or wood-frame construction. In large facilities in the United States, metal and tilt-up concrete construction is used more commonly than wood-frame construction or concrete blocks.

Walls are insulated with fiberglass batts, rigid urethane foam boards, or sprayed-on urethane foam. Batt and board insulation must be protected with a vapor barrier on the warm side. Properly applied sprayed-on foam can be moisture-proof. Exposed foam insulation must be coated with a fire-retardant barrier. Some storages combine insulation types; for example, the interior can be sprayed with foam to form a vapor barrier, and then covered with batt or board insulation for ease of sanitation and fire protection. If controlled-atmosphere techniques are used in the storage facility, the vapor barrier may also serve as a gas barrier, and special precautions must be taken to ensure a gastight seal.

Total wall insulation level, measured in heat resistance units (RSI in units of $m^2 \cdot °K/W$ or R in units of $ft^2 \cdot °F \cdot h/Btu$), is often in the range of RSI 3.5 to 7.0 (R20–R40). Ceilings can be insulated with rigid board or foam materials, or they can be suspended below the roof and insulated with loose fill or batts. Ceiling insulation of at least RSI 10.6 (R60) is common in new construction. In general, it is advisable to build with more insulation than utility costs may currently warrant because energy costs are likely to increase—and it is much less expensive to install insulation during initial construction than it is to add insulation after construction is completed.

Reducing electricity costs

A number of options are available to reduce electricity consumption in cold storages (Prakash and Singh 2008). Proper selection and control of compressors, as well as operating evaporators at the highest possible temperatures, have already been mentioned as effective conservation methods. Adding condenser coil surface area allows the refrigerant fluid to be cooled to lower temperatures and reduces compressor electricity use. Condenser fans should be controlled to minimize airflow through the condenser, based on refrigeration load and outside air temperature and humidity. The approaches mentioned above are usually implemented as a package and can reduce electricity use by 25 to 40 percent.

Sunlight falling on walls or roofs can dramatically increase the effective outside air temperature, increasing heat flow into a storage facility. (The effective outside air temperature is the normal air temperature plus a factor to account for the sun shining on a surface.) Table 2.2 shows the relationship between wall or roof orientation, color, and effective outside air temperature. A dark, flat roof can be 38°C (69°F) warmer than the outside air temperature. Painting a dark, south-facing wall a light color can reduce the effective wall temperature by 9°C (16°F). The walls and roof of a cold-storage facility should be painted a light color or shaded from the direct sun. High-reflectivity coatings for walls and roofs are now available and

Table 2.2. Effective outside air temperature when direct sunlight strikes various building surfaces in Fresno, California (latitude 40°N) on July 21

		Effective air temperature	
Surface	**Color**	**°C**	**°F**
ambient air	—	34	93
horizontal roof	light	51	124
horizontal roof	dark	72	162
south wall	light	42	108
south wall	dark	51	124
west wall	light	51	124
west wall	dark	71	160

Source: Adapted from ASHRAE 1993 (see especially chapter 20).

reduce solar heat gain even more than light-colored paints.

Reducing heat input to the refrigerated space is a major method of reducing electricity use. Fan use, and the heat that fans produce, can be minimized with speed control or fan cycling. Use efficient lighting sources like high-bay fluorescent or LED fixtures designed to operate at low temperatures. Use minimum lighting levels, supplemented with task lighting on forklifts. Install timers that shut down most of the fixtures when the storage room has been unoccupied for 5 minutes. Lighting retrofits often pay for themselves in less than a season or two of electricity savings. Use plastic flap doors, air curtains, or fast-acting doors to reduce infiltration of warm outside air during loading and unloading. Seal around openings for pipes and electrical conduits. Loading docks for outbound produce should be fitted with flexible bumpers that provide an airtight seal between the dock and a truck. Finally, consolidate produce into the minimum number of rooms and design the refrigeration system so that unused rooms can be shut down.

Small-scale refrigerated storage

Cold rooms for small-scale operations can be purchased from commercial suppliers, built by facility operators themselves, or made from used equipment for refrigerated transportation such as railcars, marine containers, or highway vans. The choice of which system to use is based on cost and availability of equipment in the local area and the amount of time available to invest in cold-room installation.

Commercial cold rooms

Cold rooms for retail stores and wholesale facilities are usually made from structural insulated panels. These are factory-built panels of expanded polystyrene or urethane foam board with metal or plastic outside surfaces. They are trucked to the site and quickly assembled with built-in connectors. They are available for installation outside or within a building. Used, panelized cold rooms can be a very economical option for small operations.

Railcars

Refrigerated railcars in the United States are very sturdy and well insulated. In older-generation equipment, the refrigeration is powered by an electric motor that is powered in turn by a diesel generator. The generator set can be removed from the car and salvaged; the refrigeration can be connected to the operator's electric utility. Newer railcars are cooled with the same units used for refrigerated highway trailers. The ceiling in the cars is 2.8 meters (9 ft 4 in), which limits the height that produce can be stacked. The most significant problem in using railcars for cold storage, as well as the greatest cost, is getting them from the railroad to the site.

Highway vans

The unique advantage of using highway vans as cold storage is that they are portable if the wheels are left on. The refrigeration system is powered by a diesel engine. This can be a benefit if utility electricity is not available at the site. In some areas, it may be less expensive to operate the refrigeration unit if it is converted to operate with an electric motor, but this approach adds a considerable conversion cost to the project. Highway vans are built to be as light as possible to maximize the load weight they can carry; this often means that used vans are in fairly poor condition when they become available on the used market. Their insulation, which is limited to begin with, may have deteriorated, and their doors may be poorly sealed, permitting excess air leakage. Also, old vans often have fairly small refrigeration fans that may not provide adequate air circulation.

Transport containers

Used containers are available in standard lengths of 6.1, 12.2, and 13.7 meters (20, 40, and 45 ft). Their built-in refrigeration units are powered with 220- or 440-volt, three-phase electricity, and they can be plugged directly into utility power. They are usually well-built and have deep T-beam floors and sufficient fan capacity to provide good air circulation; in fact, air circulation is good enough to allow adequate room cooling.

A disadvantage of all transport vehicles is that their refrigeration systems are usually not designed to produce high relative humidity. When produce dries due to low humidity, it causes weight loss and poor quality. This is a problem particularly if the cold room is to be used for long-term storage. Nearly the only way to reduce the drying is to pack the produce in plastic bags.

Refrigerated transport vehicles rarely have enough refrigeration capacity to cool produce rapidly. If rapid cooling is needed, extra capacity must be added. Moreover, transportation vehicles are too narrow for the frequent produce movement needed in a cooling facility. A separate, self-constructed room is much more convenient for cooling operations.

Self-constructed cold rooms

For many producers, a self-built cold room is the least expensive option (see Thompson and Spinoglio 1996 for more details). Self-built cold rooms usually have a concrete floor. The walls and roof are wood-frame construction insulated with fiberglass batts. Care must be taken to install a tight vapor barrier on the warm side of the insulation. Refrigeration is provided by a small commercial refrigeration system or a room air conditioner with an add-on controller.

Controlled-atmosphere storage

Controlled-atmosphere storage uses oxygen (O_2) and carbon dioxide (CO_2) concentrations different from those found in air (78% N_2, 21% O_2, 0.9% argon, 0.04% CO_2). Oxygen levels are usually decreased below 2 percent, and carbon dioxide levels are usually increased above 5 percent. These levels slow ripening, stop the development of some storage disorders such as scald in apples, slow the growth of decay organisms, and can minimize the effect of ethylene. All these effects increase storage life of fresh produce compared with refrigerated-air storage alone. More details about the potential benefits and hazards of controlled-atmosphere storage are available in *Postharvest Technology of Horticultural Crops: Atmospheric Environment.*

Simple controlled-atmosphere system

Controlled-atmosphere storage involves all the design requirements of conventional refrigerated storage—plus gastight rooms, equipment to create the desired gas concentrations, and equipment to measure and control atmospheric composition. The simplest system for obtaining gastight storage involves a plastic tent inside a conventional refrigerated-storage room (Leyte and Forney 1999). The tent is made of polyethylene sheeting, 0.0762 to 0.127 millimeters (3–5 mil) thick, supported by a wood framework. The sheeting is sealed to the concrete floor by pressing the plastic into a narrow trough and forcing tubing into the trough to keep the plastic in place. A better gas barrier at the floor can be established by laying a sheet of plastic on the floor and covering it with wood panels. A seal is established by joining the tent to the floor sheet. A fan inside the tent provides air circulation. The oxygen level is reduced initially by allowing fruit respiration to consume oxygen or by using controlled-atmosphere generators. Oxygen is kept above the prescribed minimum for the produce being stored by allowing a controlled amount of outside air to enter the tent. The carbon dioxide level is maintained by placing bags of fresh hydrated lime (calcium hydroxide) in the tent to absorb excess carbon dioxide.

Commercially available pallet covers can also be used to provide modified-atmosphere storage on a small scale.

Permanent controlled-atmosphere facilities

Permanent facilities require that the storage plant be designed specifically for controlled-atmosphere storage. Usually a controlled-atmosphere facility costs about 5 percent more to build than a conventional refrigerated-storage facility. The extra cost comes from building the storage rooms to be airtight and, in some cases, designing smaller individual rooms than those needed in conventional refrigerated storage.

Room size

Individual rooms should be sized to allow them to be filled in a short time. Many controlled-atmosphere rooms for storing apples are built to hold 1 week's harvest. If an operator wishes to use rapid controlled atmosphere, the room should be small enough to be filled in 3 days. Many facilities have several room sizes to allow for variations in incoming fruit volume, fruit varieties, and marketing strategy. In many areas of the world, rooms are sized to hold 100 to 200 tons of fruit. The state of Washington, in the United States, has a very large apple production region and supplies much of the market in North America. Controlled-atmosphere rooms in this area are sized to hold 500 to 1,000 tons of apples.

Today's controlled atmosphere storages are single-story designs, with individual rooms often tall enough to allow fruit to be stacked 10 bins high, including enough clearance between the bins and the ceiling to allow air from the evaporator coils to mix with the room air and travel easily to the far end of the room.

Walls and ceilings

Three main types of interior wall and ceiling construction are used in recently built controlled-atmosphere storages: foamed-in-place urethane over concrete or metal walls; plywood-covered stud walls with fiberglass insulation; and insulated, metal-covered panels (structural insulated panels). (See Bartsch and Blanpied 1990 for a detailed discussion of controlled-atmosphere storage design and operation.)

Urethane foam serves as insulation, a gas seal, and vapor barrier. But it is expensive, primarily because the urethane must be covered with a fire barrier. Foamed-in-place urethane can be applied directly to any type of masonry, wood, or primed metal. It should not be applied over rigid board insulation because the foam can distort the board, and board materials that are not tightly attached cause the gas seal to fail over time. Figure 2.8 shows a cross section of a typical controlled-atmosphere storage sealing system with urethane foam insulation in the walls.

Installing a plywood cover over fiberglass insulation is another option for insulating and sealing controlled-atmosphere storage. The seams between plywood sheets are sealed with an elastomeric compound, like butyl rubber, applied between the sheet and the wood framing. Sheets are separated by a gap of 3 millimeters (0.12 in) to allow for expansion. The gap is filled with the same elastomeric compound and the joint is covered with fabric and an elastomeric sealer. If regular plywood is used, the whole board must be covered with the sealer. High-density plywood does not need to be treated for gastightness. In most controlled-atmosphere storage, a vapor barrier is installed on the outside (warm side) of the insulation. An outside vapor barrier is not recommended with the plywood system because there is already a gas-tight seal on the inside of the insulation. A vapor barrier on the outside traps moisture that might get into the insulation, ruining its insulating value.

Structural insulated panels are usually a sandwich design in which 8 to 15 centimeters (3–6 in) of rigid foam insulation are covered on both sides with painted metal or plastic sheets. Panels are 1.2 to 2 meters (4–6 ft) wide and usually extend from floor to ceiling. The panels are installed inside a building shell and are held together with mechanical fasteners. Joints are sealed with materials such as polyvinylacetate

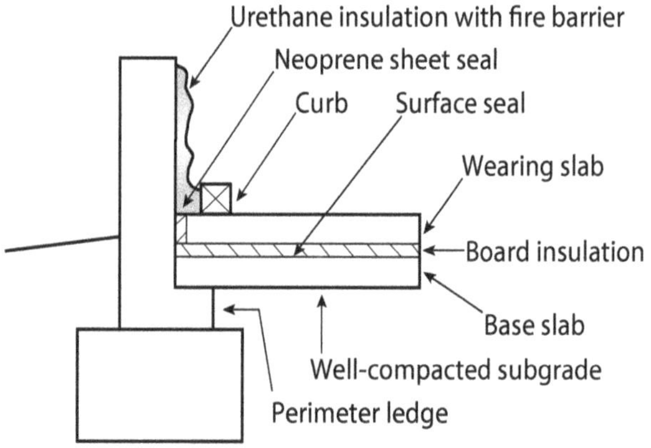

Figure 2.8. Gastight sealing technique using urethane foam insulation. Adapted from Bartsch and Blanpied 1990.

copolymer or a latex emulsion. Large gaps are backed with a nonwoven fabric.

Floors

Floors can be made gastight whether a building uses layered or single-slab designs. An insulated floor is usually a sandwich design, with board insulation placed between two layers of concrete (see fig. 2.8). With this design, a gas seal of two layers of hot-mopped asphalt roofing felt is applied to the subfloor. A single-slab floor is used if only perimeter insulation is used. In this case the floor is sealed by applying special materials such as chlorinated rubber compounds to the top surface. In both designs, the wall-to-floor seal is the area most likely to fail. Prevent the floor from moving with respect to the wall by carefully backfilling and thoroughly compacting the subgrade. The floor can be tied to the walls with rebar, or the floor can sit on a ledge, 10 centimeters (4 in) wide, built into the foundation wall.

Doors

Many door designs can be used in controlled-atmosphere facilities. In all of them the door is constructed of a solid frame that can be clamped tightly against a gasketed doorframe without warping. The frame can be covered with well-sealed plywood or metal. Some of the more expensive doors have aluminum sheets welded to an aluminum frame. The bottom of the door is usually sealed with caulking compound after the door is closed. Most doors are 2.4 to 3 meters (8–10 ft) wide and tall enough to allow a lift truck with two bins to pass through.

Each room should also have an access door, 60 by 75 centimeters (24 by 30 in), to allow workers entry to check fruit and make repairs without opening the main door. Many storages also have a clear acrylic window near the top of a wall to allow workers to inspect the fruit without entering the room. The window is usually a concave shape, allowing a view of all areas of the room.

Pressure relief

A pressure difference between the cold room and the outside can develop because of changes in weather or room temperature. This difference can damage the gas seal if it is not relieved. A water trap (fig. 2.9) can be used to allow pressures to equalize. The trap is often filled with ethylene glycol to prevent the water from evaporating. A spring-loaded or weight-loaded check valve can also be used, but both are more expensive than a water trap. Provide 10 square centimeters of vent opening for each 40 cubic meters of room volume (1 in^2 per 1,000 ft^3).

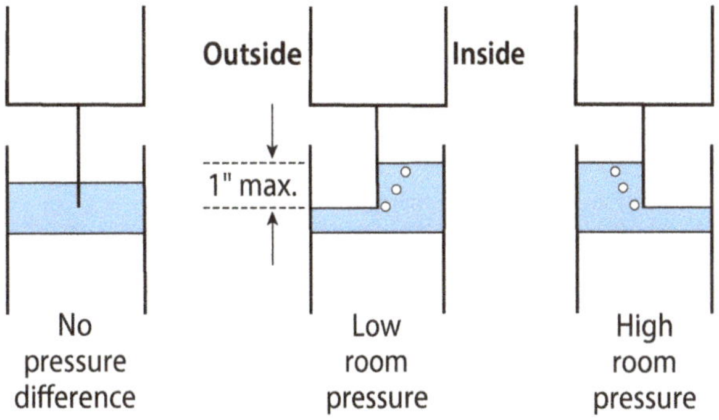

Figure 2.9. Schematic of a water-trap pressure-relief system.

Small changes in pressure can also be relieved by using breather bags. These have the advantage of capturing the gas mixture in the room and allowing it to reenter the room at a later time. Bags should have about 0.5 cubic meters of capacity per 100 cubic meters of room volume (5 ft^3/1,000 ft^3), although some are designed with several times this capacity. If a controlled-atmosphere room is so tightly sealed that air must be regularly bled into the room to maintain O_2 level, breather bags are not necessary.

Pressure test

The overall gas-tightness of a controlled-atmosphere room can be tested by pressurizing it and measuring the rate of pressure drop. After all doors and openings are sealed, a small fan is used to increase the static pressure in the room to 25 millimeters (1 in) of water column. The fan is then turned off and the room is sealed, and the tester determines how long it takes for the pressure to drop by one-half. If it takes 20 minutes or longer, the room is considered tight enough for a controlled-atmosphere system that uses hydrated lime to maintain carbon dioxide levels. At least 30 minutes are required for rooms that use carbon or water scrubbers to control carbon dioxide. Some researchers believe that rooms should be even more gas-tight than these standards, recommending 45 minutes to 2 hours for the pressure to drop by one-half. Because gas seals deteriorate, new rooms should be much tighter than prevailing standards.

If the room does not meet the desired standard, check for leaks. Leaks are most common around doors, at the wall and floor junctions, near poorly sealed penetrations, and at unsealed electrical boxes. Test for leaks by putting the room under a slight vacuum and then listen for air leaking—or spray suspect areas with soapy water and watch for bubbles. Smoke sticks or bee smokers can also be used to detect streams of outside air entering the room.

Atmospheric modification

The least expensive but slowest method of modifying the storage room atmosphere is to let the produce modify it through natural respiration. Fruit and vegetables consume oxygen and release carbon dioxide. The produce filling a sealed room eventually lowers the oxygen to the level needed for controlled-atmosphere storage. If oxygen drops too low, outside air can be added to restore it to the desired range. However, respiration causes carbon dioxide levels to rise well above required levels. Bags of hydrated lime can be used to absorb excess carbon dioxide. Lime requirements are 1 to 3 kilograms per 100 kilograms of produce (1–3 lb per 100 lb), depending on the produce being stored, storage time, surface area, and quality of the lime. Bags can be placed either in single layers in the controlled-atmosphere room or in an adjacent lime room that is connected to the controlled-atmosphere room with a fan and ductwork. Rooms should be sized to hold about 15 kilograms of lime per metric ton of produce (30 lb/ton). Lime is very effective at producing the low levels of carbon dioxide that are increasingly used in apple storage. Carbon dioxide levels can also be controlled with activated carbon adsorption systems or molecular sieves. Molecular sieves tend to use more energy than activated carbon systems.

Relying on produce respiration to remove oxygen is fairly slow, and the produce storage life can be increased if the oxygen is removed faster. Some operations purge the controlled-atmosphere room using nitrogen purchased in liquid form or produced on-site. Two common nitrogen generator systems use a molecular sieve process (pressure swing adsorption) or a semi-permeable membrane to remove oxygen.

Refrigeration equipment

Refrigeration equipment for controlled-atmosphere facilities is the same as for any other cold-storage operation. Most storages are designed to maintain 0°C (32°F) at 95 percent relative humidity. Up to 0.052 $m^3 \cdot sec^{-1} \cdot$metric ton^{-1} (100 cfm/ton) of air circulation is required for produce cooling after initial loading. Air circulation during storage should be in the range of 0.0208 to 0.0104 $m^3 \cdot sec^{-1} \cdot$metric ton^{-1} (20 to 40 cfm/ton) of maximum storage capacity. After produce has reached storage temperature and cool outside temperatures reduce refrigeration demand, evaporator fans are regularly cycled off to reduce heat load and the need for refrigeration. This reduces energy use and increases room relative humidity.

Monitoring equipment

Oxygen and carbon dioxide levels must be regularly monitored to ensure that they are within prescribed limits. Electronic gas sensors are used to measure gas concentrations. Computer-based controllers use the gas level information to operate ventilation systems and other gas control equipment to automatically maintain proper gas concentrations.

Temperature should also be monitored regularly. A minimum of two calibrated dial thermometers should be installed in each room. One should be near eye level, with the dial located on the outside of the room. The other should be above the commodity and should be readable through an observation window near the ceiling. Electronic thermometers allow easier observation of room temperature, and the data can easily be printed out for a permanent record of operating conditions. Most new storages use four temperature sensors (or more) per room. Some sensors are placed into bins to monitor commodity temperatures.

Measuring produce temperatures

Bimetallic thermometers are inexpensive devices for measuring produce temperature. They can usually be recalibrated with an adjusting nut located under the thermometer's dial. Bimetallic thermometers can be purchased with dials large enough to read to the nearest degree, but because they tend to respond slowly to produce temperature, operators need to determine the response time of their equipment.

Electronic thermometers (thermocouple or thermistor units) can be purchased with very thin, strong probes. The resolution of many units is one-tenth of a degree. The response time of each unit must be determined before prescribing the protocol for measuring produce temperature. Electronic thermometers can be purchased with lighted readouts for operation in dark areas. Most units can display both Fahrenheit and Celsius temperature scales.

Electronic thermometers fitted with long lead wires allow measurement of coolant temperatures in inaccessible locations. Thermocouple wire costs about 10 cents per foot, and a readout device costs as little as $100. Thermistor sensors, which are more accurate than thermocouples, cost more but allow the use of less-expensive copper lead wires. Temperature sensors enabled for Wi-Fi or Bluetooth transmit data to handheld readout devices, phones, or other logging devices, without the need for wires to transmit data.

Infrared thermometers give instantaneous responses, but they measure the surface temperature of the material in front of them. They do not "see" through packaging material. Surface temperature may represent produce temperature if the produce has been in a constant temperature environment for many hours, as is the case for produce in long-term storage. During and shortly after initial cooldown, the produce's surface temperature is usually the coldest portion of the produce and not a good indicator of average temperature. The temperature reading also depends on a property of the surface called emissivity (the effectiveness of a surface to emit radiant heat). Some units have an emissivity adjustment to account for emissivity differences between plant material, packaging materials, or other surfaces. Units without emissivity adjustment are usually calibrated for high emissivities and are well suited for perishables, but may not be accurate for painted or bare metal surfaces. These units work well for rapidly surveying the temperature of produce coming out of refrigerated vehicles.

Cold produce may absorb enough heat from a thermometer to warm the flesh near the probe, resulting in an erroneously high temperature reading. This error can be prevented by probing an item once to bring the thermometer near the produce temperature and then reinserting it in another location to take the final reading. Double probing also decreases response time of the temperature sensor. During cooling, the warmest part of a produce item will be near its center. Probes are usually inserted in the center of a produce item when the effectiveness of a cooler is monitored. Some probes, particularly bimetal thermometers, have an immersion depth marked on the stem. All of the probe, from the tip to the immersion mark, must be within the produce for accurate temperature measurement. Punctured produce should be discarded if the puncture damage is in a portion of the produce that will be consumed.

Management decisions for cooling operations are usually based on the temperature of the warmest produce in the operation. Operators need to survey produce temperature thoroughly in different positions within boxes, in different box positions on pallets, and in different pallet or bin locations in a cooler or storage to determine where the warmest produce is located. After this, temperature monitoring can be focused on the locations where produce is known to be warmest. For example, in tunnel-type forced-air coolers, the warmest produce is usually in the pallets farthest from the fan and in boxes near the air-return channel. In forced-air coolers with airflow reversal, the warmest produce is in the center of the pallet load.

Storage operations are usually evaluated on the basis of both average and extreme produce temperatures. Many measurements should be taken within a batch—top to bottom, side to side, and end to end. Average temperature indicates the overall calibration of the refrigeration and control system. Extreme temperatures (high or low) point to local problems caused by factors such as poor airflow, inadequate insulation, poor mixing of air from evaporators with storage room air, and excess air infiltration.

All temperature-measurement equipment should be calibrated at least once a year. Accuracy can be checked with an ice-water bath, which has a temperature of 0.0°C (32.0°F). Infrared thermometers can also be calibrated by pointing them at the water in a stirred ice bath. The ice-bath container should be filled with crushed ice and then filled with clean water. Continuously stir the mixture with the temperature sensor until the sensor reaches a constant temperature. For use with produce held near its freezing temperature, temperature sensors must be calibrated to the nearest one-tenth of a degree. Many instruments allow adjustments for recalibration. If adjustment is not possible, purchase a new instrument or write the error on the thermometer. The calibration date should be marked on each sensor.

Temperature measurement is time-consuming, but it is a valuable management tool for evaluating the efficiency of the operation and must be done routinely to ensure that equipment is operating properly and produce temperature goals have been achieved.

References

Anderson, T. N., M. Duke, and J. K. Carson. 2013. Performance of an unglazed solar collector for radiant cooling. Proceedings of the Australian Solar Cooling 2013 Conference:1–9. https://openrepository.aut.ac.nz/handle/10292/5651

(ASHRAE) American Society of Heating, Refrigeration, and Air-Conditioning Engineers. 2021. ASHRAE handbook: Fundamentals. Atlanta: American Society of Heating, Refrigeration, and Air-Conditioning Engineers.

Bartsch, J. A., and G. D. Blanpied. 1990. Refrigeration and controlled atmosphere storage for horticultural crops. Northeast Regional Agricultural Engineering Service. NRAES-22. https://archive.lib.msu.edu/DMC/Ag.%20Ext.%202007-Chelsie/PDF/e1914-ND.pdf

Kitinoja, L., and J. F. Thompson. 2010. Pre-cooling systems for small-scale producers. Stewart Postharvest Review 2(2):1–14.

Koca, R. W., and M. L. Hellickson. 1993. Energy savings in evaporator fan-cycled apple storages. Applied Engineering in Agriculture 9(6):553–560. https://doi.org/10.13031/2013.26021

Leyte, J. C., and C. F. Forney. 1999. Controlled atmosphere tents for storing fresh commodities in conventional refrigerated rooms. HortTechnology 9(4):672–675. https://doi.org/10.21273/HORTTECH.9.4.672

Prakash, B., and R. P. Singh. 2008. Energy benchmarking of warehouses for frozen foods. Davis: Public Interest Energy Research (PIER) Program Contract and Research Project Reports; California Energy Commission, PIER Program.

Prussia, S. E., and R. L. Shewfelt. 1985. Ventilation cooling of bulk shipments of shelled southern peas. Transactions of the ASAE 28(5):1704–1708. https://doi.org/10.13031/2013.32503

Stoecker, W. F. 1998. Industrial refrigeration handbook. New York: McGraw-Hill Education.

Thompson, J. F., and M. Spinoglio. 1996. Small-scale cold rooms for perishable commodities. Oakland: UC Agriculture and Natural Resources Publication 21449. http://postharvest.ucdavis.edu/Library/Publications/?uid=535&ds=807

Chapter 3
Safety and health hazards

Farzaneh Khorsandi

Initial cooling and cold storage operations, like all industrial operations, have the potential for accidents. According to the U.S. Bureau of Labor Statistics, in 2019 the rate of occupational injuries and illness in the "refrigerated warehousing and storage" sector in the United States was about 5.5 per 100 workers. This means that the "refrigerated warehousing and storage" industry is a relatively dangerous industry, compared with the "agriculture, forestry, fishing, and hunting" sector, which is among the most dangerous industries in the United States, with an occupational injuries and illness rate of 5 reports per 100 workers (BLS 2022). Refrigerated facilities must be carefully designed and operated to maximize worker health and safety.

Hazards in cooling systems

Working in cooling systems and industrial refrigeration involves three main types of risk: physical, biological, and chemical. The physical hazards in this industry, rather than being specific to refrigeration, are common to many work situations. Below, the physical hazards are discussed briefly, while biological risks (specifically *Legionella* bacteria) and chemical exposure (specifically anhydrous ammonia) are addressed in more detail.

Physical hazards

A physical hazard is an environmental factor that can hurt the body—mainly by physical contact but in some cases, as with vibration and noise, without necessarily touching the body. Some of the physical hazards associated with cooling systems are listed in table 3.1. A comprehensive physical safety training is available through the Occupational Safety and Health Administration's 10-Hour General Industry (Agriculture) course.

Biological hazards

The biological hazards, also known as biohazards, can be caused by organisms such as viruses, bacteria, insects, plants, animals, and humans. These biological hazards can cause a variety of human health effects, ranging from

Table 3.1. Physical, biological, and chemical hazards in cooling systems

Physical hazards	Biological hazards	Chemical hazards
• occupational injuries (sharp objects, hot/cold surfaces) • ergonomics • struck by object • falling objects • flying objects • slips, trips, and falls • compressed gas • explosion • moving vehicles (e.g., forklifts, powered industrial trucks) • contact with machinery • confined space • asphyxiation • electrical risks, overhead and underground electric lines • workplace violence	• fungi, bacteria (*Legionella* bacteria) • sensitizers/ allergens (dermal) • sensitizers/ allergens (respiratory)	• anhydrous ammonia • disinfectants • cleaning chemicals

minor skin irritation to life-threatening disease. Some of the biological hazards in cooling systems are listed in table 3.1. Also, the inadequate refrigeration of chilled foods leads to food spoilage, which is a source of biological hazards. The microbiological consequences of insufficient facility maintenance include Legionnaires' disease.

One of the common biological hazards when maintaining a cooling tower is respiratory infection by *Legionella* bacteria. These bacteria can grow in fresh water sources such as groundwater, rivers, and human-made water systems. Legionnaires' disease can develop after a worker inhales mist from contaminated water. The growth of *Legionella* can be controlled in the following manner:

- A water treatment program should be implemented that uses an appropriate combination of a locally permitted bactericide, corrosion and scale inhibitors, dispersants, surfactants, or filters. The cooling tower water should be continuously circulated. Organic matter, corrosion in the cooling tower structure, and water-borne bacteria can shield *Legionella* from bactericides and must all be controlled.
- Aspects of water chemistry, such as pH and chlorine concentration, can change over time and should be monitored regularly to maintain microbial control. Indicators of poor water quality management are the presence of green or brown algae; an accumulation of soil, trash, or other debris; a foul smell; or corrosion or rusty color. *Legionella* growth is correlated with high levels of other bacteria in cooling tower water.
- If the water circulation system is shut down for more than 3 days, the water should be drained, or locally permitted biocides should be added to the water because *Legionella* bacteria can proliferate in standing water.
- System maintenance should be conducted at least twice per year because damaged components such as droplet eliminators or clogged spray nozzles could increase the risk of exposure to mist that may be contaminated with bacteria.
- The water temperature should be kept below 20°C (68°F) because *Legionella* doesn't grow well below this temperature.
- The system should be maintained and cleaned regularly because biocides and other inhibitors could be inactivated by trash and debris. Scale, slime, corrosion, or sediment can clog spray nozzles, enhancing bacterial growth (Nowacki and Palm 2003).

More information about *Legionella* bacteria can be found in sources such as Ontario Ministry of Labour (2005), (IHSA n.d., Health and Safety Executive (2013), and ASHRAE (2000, 17)—all listed in the references below.

Chemical hazard: Anhydrous ammonia

Anhydrous ammonia (NH_3) is a commonly used refrigerant in vapor-compression refrigeration. Ammonia release accidents can cause fatalities and injuries to employees, people in surrounding communities, and emergency response personnel.

The word *anhydrous* means "without water." When used as a refrigerant it is usually just called ammonia, and this is the term used below.

Chemical properties

Ammonia is an alkaline chemical that is hygroscopic, readily reacting with water to form ammonium hydroxide (NH_4OH), a strong base whose pH can be between 11 and 12.5. Released ammonia quickly absorbs moisture from the air and creates a dense, visible white cloud of ammonium hydroxide that travels along the ground until the cloud's temperature equalizes with the air temperature. The visible cloud diffuses to invisibility as it expands out into the environment and warms. Also, ammonia is attracted to moist areas of the body like the eyes, nose, mouth, lungs, throat, and skin and can cause chemical burns (New York State Department of Health 2004).

At low concentrations, ammonia is nonflammable. At higher concentrations (16–25% by weight)—which are more likely to occur when ammonia is released in a confined space—ammonia can be ignited by something as common as the arc from an electric switch, though ammonia is difficult to burn.

Physical properties

At room temperature and normal atmospheric pressure, anhydrous (or pure) ammonia is a colorless gas with a strong, pungent odor, even at low concentrations. Pure ammonia gas is lighter than air but sinks to the ground if moisture is present. In atmospheric-pressure conditions, ammonia boils at a temperature of –33°C (–27°F) and melts at –77°C (–171°F). Gaseous ammonia can be easily compressed to form liquid ammonia. The release of pressure in liquid ammonia systems causes rapid evaporation, which cools the liquid to its boiling point. Liquid ammonia is less dense than water.

Health effects

Due to its caustic nature, ammonia is hazardous if absorbed into the body. Exposure to ammonia particularly affects the eyes, lungs, and skin (ATSDR 2014). It rapidly penetrates the eyes, causing swelling and burns. Without proper treatment, permanent eye damage or blindness may occur. The inhalation of ammonia causes the formation of ammonium hydroxide in the respiratory system. At low concentrations, coughing and throat irritation occur. At high concentrations, exposure causes swelling and burns, obstructing airways and causing fluid build-up in the lungs. Airway damage may occur, which leads to respiratory failure. Contact with liquid ammonia causes three types of injuries concurrently. Ammonia has a very low temperature at ambient pressure and can cause frostbite. Liquid ammonia in contact with flesh absorbs water, causing dehydration injury. The reaction between ammonia and the body's water causes a chemical injury due to ammonia's high pH and the heat of the reaction. The low temperatures may also freeze clothing to the skin, which tears if clothing is improperly removed.

Eyes and skin exposed to ammonia should be continuously flushed with clean water for 15 minutes. Never apply water to a pool of liquid ammonia; the result will be a chemical reaction and a dense vapor release.

The range of exposure and effects

The range of ammonia exposure and effects is shown in figure 3.1. The values are listed in parts per million. Humans can smell the presence of ammonia at one part per million (Memarzadeh and Manning 2005). The U.S. Occupational Safety and Health Administration (OSHA) Recommended Exposure Limit for ammonia is 25 parts per million over 8 hours. The Federal OSHA Permissible Exposure Limit for ammonia is 50 parts per million, averaged over 8 hours. According to the American Conference of Governmental Industrial Hygienists, the Short-Term Exposure Limit is 35 parts per million (a time-weighted average) during any 15-minute period in the day. Thus, workers should not be exposed to more than this amount over any 15-minute period (Gay and Knowlton 2005). According to the National Institute for Occupational Safety and Health

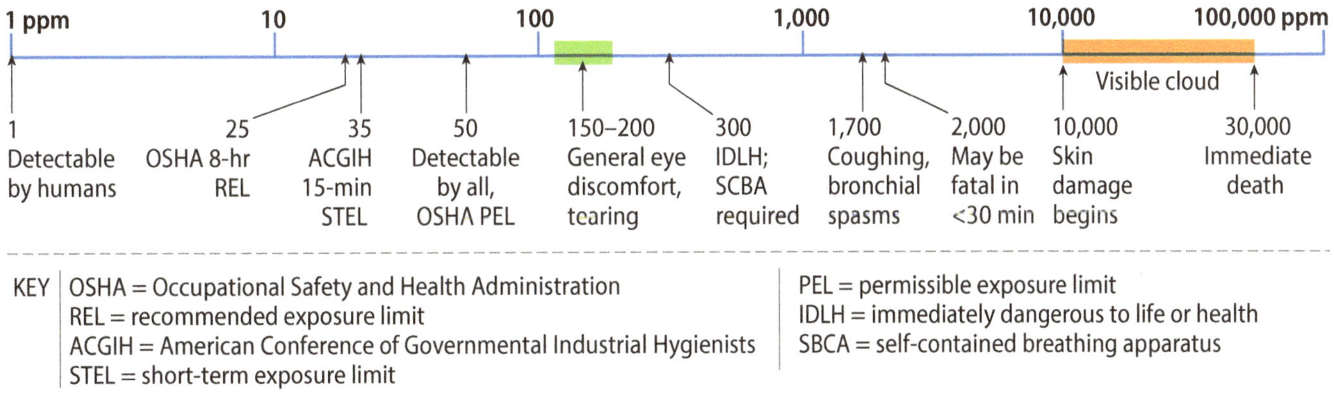

Figure 3.1. Ammonia exposure limits and effects on humans.

(NIOSH), the concentration of ammonia immediately dangerous for life or health is 300 to 500 parts per million for 30 to 60 minutes as a maximum short-exposure tolerance (NIOSH 1988). According to the hazard guide of the Ammonia Safety and Training Institute (ASTI 2018), 2,500 to 5,000 parts per million of ammonia exposure can cause edema, asphyxia, and skin sting, and may be fatal in 10 to 30 minutes, or less. Inhalation of 10,000 to 30,000 parts per million of ammonia vapor can cause immediate death by asphyxiation and pulmonary edema; it may also cause skin damage, redness, sloughing, and blistering.

Training

Operators of refrigerated facilities should be qualified and fully trained to work in the facility. Training includes awareness of hazards associated with ammonia release and the rapid reporting of an ammonia release. Several organizations provide training and education for operators, including the Refrigerating Engineers and Technicians Association, the Industrial Refrigeration Consortium at the University of Wisconsin-Madison, and the International Institute of Ammonia Refrigeration (IIAR). Operators should take refresher training at least every 3 years to ensure that they understand and follow operating procedures. In the case of rotating jobs, operators should be trained before starting a new job. Operators should be retrained after significant changes in the operation procedures in the facility, system damage, or a potential accident.

Personal protective equipment

Personal protective equipment is an ensemble designed to be worn by individuals for protection against possible hazardous threats such as ammonia release and asphyxiating vapor. People who work with ammonia are subject to potential overexposure and the proper safety equipment should be provided for them (fig. 3.2). Personal protective equipment includes four categories (A, B, C, D) depending on the level of protection it affords. The Ammonia Safety and Training Institute recommends the use of Level

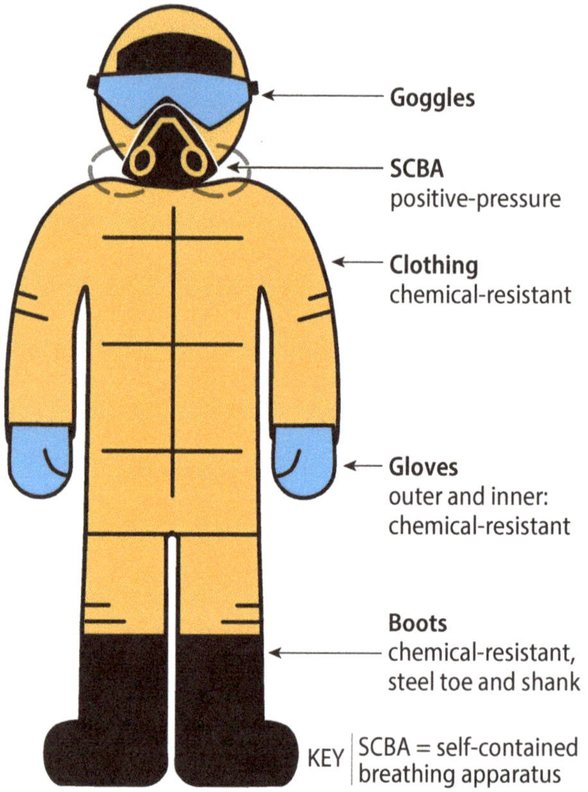

Figure 3.2. Level B personal protective equipment.

B personal protective equipment in less than 15,000 parts per million of ammonia vapor. This limitation has been increased from the limit of 5,000 parts per million that OSHA established in 1991 (Smith 2014). Level B personal protective equipment should be worn when the highest level of respiratory protection is required, but a lower level for eye and skin protection is needed (CHEMM 2021). Eyewash/shower stations must be placed in the facility so workers can reach them within 10 seconds of exposure. IIAR calls for eyewash/shower stations to be located inside and outside refrigeration machinery rooms. The ensemble of Level B personal protective equipment includes the key components described in figure 3.2 (ASTI 2018; CCFPCCD 2009).

OSHA recommends (OSHA 2004) that "If impervious clothing is not available, wear cotton, which is the preferred fabric for work clothing because it is more alkali-resistant than wool and is more comfortable than all-synthetic fibers. Clothes, especially gloves, should also be insulated to prevent freezing of the skin."

Facility safety requirements

The Emergency Response and Community Right-To-Know Act and Title III of the Superfund Amendments and Reauthorization Act require facilities that handle hazardous chemicals to have plans for accident prevention and emergency response to accidents. In addition, Congress has enacted laws to keep employees, emergency responders, and surrounding communities safe. Several agencies—such as OSHA, the Environmental Protection Agency, and the Department of Homeland Security—enforce laws and provide regulations to help make operators of anhydrous ammonia refrigeration facilities aware of their responsibility to prevent accidental ammonia releases and manage releases safely.

Each facility must consider federal, state, and local requirements. For example, OSHA requires ammonia refrigeration facilities with 10,000 pounds of ammonia (approximately 2,000 gal) or less to comply with the General Duty Clause of the Occupational Safety and Health Act of 1970—and refrigeration facilities with 10,000 pounds of ammonia or more to follow the requirements of the OSHA Process Safety Management Standard. The EPA has published a document that educates anhydrous ammonia refrigeration operators about safely managing their facilities and fulfilling their responsibility to prevent accidental ammonia releases (EPA 2015).

Preventing accidents

The majority of chemical accidents can be avoided through operator training and well-designed operating procedures. Operators should be aware of the hazards associated with ammonia release and safety measures that can prevent accidents. Also, system operation should be controlled appropriately, regular maintenance should be performed, and the system should be regularly inspected. Before employees begin work, they should be trained so that they understand facility operations and maintenance procedures. Some recommendations for preventing accidents appear below:

- **Remove oil** as needed. The operator performing oil removal should be equipped with the personal protective equipment required at the plant and a second operator should be present. The oil pot should not be frosted during oil draining.
- **Protect ammonia equipment, vessels, and piping** from falling objects by using horizontal and vertical structural members. Concrete curbs, bollards, and barriers should be installed to prevent impact by forklifts, trucks, and other maintenance vehicles.
- **Install an ammonia detector system** to monitor rooms where leaks could occur and areas where workers are not present 24 hours a day, 7 days a week. When the system detects ammonia, a notification should be sent to a monitored location. *Monitored location* means a pager on the person of a staff member or responsible party. A pager is recommended rather than a cell phone for several reasons:
  - Most cell phones use touch screens and require the operator to remove a glove to work with the phone.
  - A pager dramatically increases the chances of messages getting through.
  - During emergencies, hospital staff can reach hundreds of people with a single page alert, without creating large group texts. Paging systems also have up to seven times the power of cellular networks, translating into better signal penetration in buildings and more reliable message delivery.
- **Install ventilation systems** to remove ammonia from the facility in case of an accidental release. There should be a mechanical ventilation system that runs continuously in the machinery rooms of the facility. Some facilities install remote manual switches to activate the ventilation system in case of an ammonia release. The switch should be close to the compressor room. Some facilities have installed ammonia detectors that automatically activate ventilation systems.

- **Install a check valve close to the main control valve** to stop ammonia release in case the main control valve is inoperable.
- **Install a remotely controlled solenoid valve on the king valve line** near the receiver vessel to control the king valve line in case of an ammonia release. The manual control switch should be located outside the refrigeration equipment room and should be recognizable by personnel and emergency responders.
- **Apply color coding, signs, and labeling** to reduce the chance of error in the operation of the facility (ASME 2020). A typical pipe label for ammonia is described by figure 3.3. From the left, the first section of the label includes abbreviations to identify system functions (*LTRS* means low-temperature liquid suction). The second section indicates the physical state of the refrigerant (liquid = yellow or orange; vapor = blue; liquid and vapor = yellow or orange and blue). The third section is the name of the gas, *ammonia*, in black letters on a yellow or orange background. (ASME changed the background color from yellow to orange in about 2014. It is acceptable to have pipe markers with yellow or orange backgrounds at the same facility.) The fourth section shows the refrigerant pressure (low or high). The fifth section is an arrow indicating the direction of ammonia flow in the pipe.
- **Label valves, pipes, king valves, emergency valves, and other components** uniquely and appropriately. Ammonia placards such as the National Fire Protection Association (NFPA) 704 anhydrous ammonia diamond with fire hazard indicator 3 should be installed in areas where ammonia refrigerant is used or stored indoors (fig. 3.4). The NFPA 704 anhydrous ammonia diamond with fire hazard indicator 1 should be installed for outdoor use or storage (fig. 3.5).

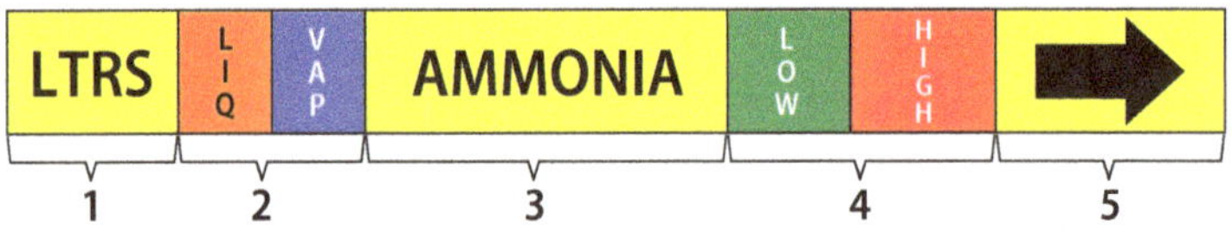

Figure 3.3. The five components of an ammonia pipe marker.

Figure 3.4. National Fire Protection Association hazard diamond for indoor use of anhydrous ammonia refrigerant.

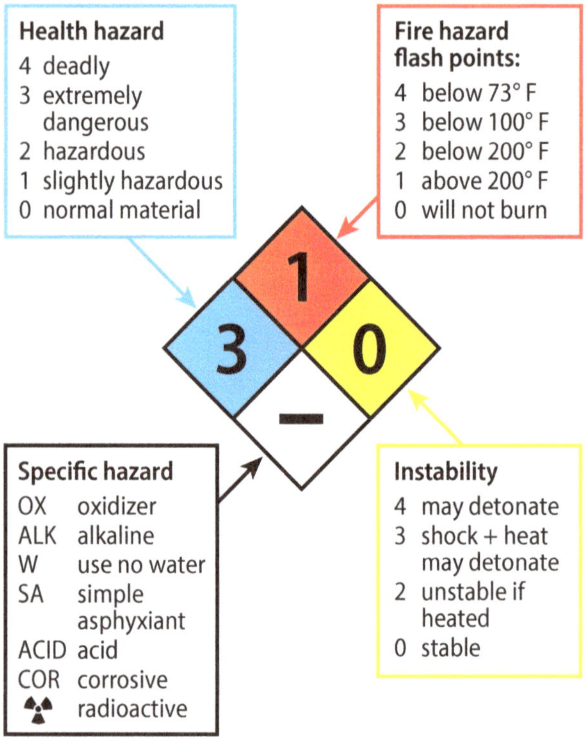

Figure 3.5. National Fire Protection Association 704 anhydrous ammonia diamond for outdoor uses of ammonia.

- **Monitor the operation parameters of refrigeration systems** to prevent accidents. The operation parameters—such as refrigerant volume and temperature, equipment vibration, lubrication levels, and pressures—should be measured and logged at least once per shift. The operator should pay attention to changing trends in the parameters because they can be indicators of system problems. Record ammonia additions to the system and monitor ammonia distribution in the system. Increasing ammonia additions and atypical distribution of ammonia are indicators of possible ammonia releases.

- **Ensure that components in refrigerated facilities** aren't made of brass, copper, or galvanized surfaces and materials. Ammonia is corrosive to these materials.
- **Keep a current piping and instrumentation diagram**, which describes all system components, in the facility. Also, operation procedures, maintenance checklists, daily logs, the facility management plan, and emergency response materials should be maintained in the facility.
- **Conduct process hazard analyses routinely** (at least every 5 years) to evaluate, identify, and control hazards that exist in the facility. Process hazard analyses aim to identify possible ammonia accidents that might affect workers, first responders, surrounding communities, and the environment.
- **Conduct regular system inspections** to prevent equipment failure. IIAR developed Standard 6 in 2019, which describes inspection, testing, and maintenance of closed-loop ammonia refrigeration systems. The inspection can be conducted through visual observation, leak testing, vibration testing, and thermal imaging. Visual observation involves checking system components for corrosion and checking insulation for its integrity. The operator should take photos of all visual signs of equipment problems and take action to remedy suspected problems. Leak testing includes using sulfur sticks, litmus paper, or a portable ammonia meter to detect leaks. Also, detectable ammonia odors should be reported and action should be taken to eliminate leaks. Vibration testing is usually conducted on rotating equipment such as pumps and compressors. Vibration levels should be logged and analyzed to determine if an abnormal trend is developing, and the machine should be repaired as necessary. Excessive vibration indicates equipment is not performing as designed, which can lead to equipment damage and possibly a release of ammonia. Thermal (infrared) imaging can detect problems at an early stage, even before other inspection methods can detect problems. A temperature difference such as a hot spot is a sign of electrical resistance and high friction, which in turn are signs of possible malfunctions in electrical circuits and mechanical systems.

Responding to an accident

OSHA requires that facilities maintain an emergency action plan (OSHA 2002), which can help organize workers and supervisors during an emergency accident. At a minimum, the plan should

- list where to report the emergency
- describe emergency escape routes, evacuation procedures, and assignments
- identify workers who should remain to operate critical plant equipment before escape
- list the responsibilities and procedures for all employees and the employer after an evacuation procedure has been completed
- describe the duties and responsibilities of first responders
- list the contact information of persons who can provide further information or explanation of duties

References

(ASHRAE) American Society of Heating, Refrigerating, and Air-Conditioning Engineers. 2000. Minimizing the risk of Legionellosis associated with building water systems. Atlanta, GA: American Society of Heating, Refrigeration, and Air-Conditioning Engineers.

(ASME) American Society of Mechanical Engineers. 2020. Scheme for the identification of piping systems. A13.1. https://www.asme.org/codes-standards/find-codes-standards/a13-1-scheme-identification-piping-systems

(ASTI) Ammonia Safety and Training Institute. 2018. All-hazard response with emphasis on ammonia. Watsonville, CA: Ammonia Safety and Training Institute.

(ATSDR) Agency for Toxic Substances and Disease Registry. 2014. Medical management guidelines for ammonia (NH3). https://nasdonline.org/7433/d002592/medical-management-guidelines-for-ammonia-nh3.html

(BLS) Bureau of Labor Statistics. 2022. Injuries, illnesses, and fatalities. Table 1: Incidence rates of nonfatal occupational injuries and illnesses by industry and case types, 2020. https://www.bls.gov/web/osh/summ1_00.htm#soii_n17_as_t1.f.5

(CCFPCCD) Connecticut Commission on Fire Prevention and Control Certification Division. 2009. Donning and doffing procedure: Level "B" suit and SCBA. https://portal.ct.gov/-/media/CFPC/_old_files/DonningandDoffingLevelBpdf.pdf?la=en

(CHEMM) Chemical Hazards Emergency Medical Management. 2021. Personal protective equipment (PPE).

(EPA) Environmental Protection Agency. 2015. Anhydrous ammonia at refrigeration facilities under scrutiny by U.S. EPA. Enforcement Alert 14(2). https://www.epa.gov/sites/default/files/2015-02/documents/112renforcementalert.pdf

Gay, S. W., and K. F. Knowlton. 2005. Ammonia emissions and animal agriculture. Blacksburg, VA: Virginia Cooperative Extension Publication 442-110. https://citeseerx.ist.psu.edu/viewdoc/download?doi=10.1.1.364.4440&rep=rep1&type=pdf

Health and Safety Executive. 2013. Legionnaires' disease: The control of legionella bacteria in water systems. 3rd ed. https://www.hse.gov.uk/pubns/priced/l8.pdf

Memarzadeh, F., and A. Manning. 2005. Control of ammonia production in animal research facilities through ventilation system design. ASHRAE Transactions 111:203–212.

New York State Department of Health. 2004. The facts about ammonia: Technical information. https://www.health.ny.gov/environmental/emergency/chemical_terrorism/docs/ammonia_tech.pdf

(NIOSH) The National Institute for Occupational Safety and Health. 1988. Ammonia. https://www.cdc.gov/niosh/pel88/7664-41.html

Nowacki, J., and B. Palm. 2003. Refrigeration accident analysis. Paper presented at the 2003 International Congress of Refrigeration, Washington, DC. Paper ICR0580.

Ontario Ministry of Labour. 2005. Workplace health and safety guideline—Legionella. www.labour.gov.on.ca/english/hs/pubs/ib_legionella.php

(OSHA) Occupational Safety and Health Administration. 2002. Emergency action plans. 1910.38. https://www.osha.gov/laws-regs/regulations/standardnumber/1910/1910.38

———. 2004. Ammonia refrigeration: Emergency response. https://www.osha.gov/etools/ammonia-refrigeration/emergency-response#accordion-73090-collapse2

Smith, G. 2014. Establishing safe criteria for wearing level B chemical vapor protective clothing while working in atmospheres of less than 15,000 ppm of ammonia vapor. Paper presented at the 36th Annual Meeting of the International Institute of Ammonia Refrigeration, Mar. 23–26, Nashville, TN. http://web.iiar.org/membersonly/PDF/TC/T435.pdf

Chapter 4
Food safety

Irwin R. Donis-González and Erin DiCaprio

Each year in the United States, human pathogens cause an estimated 48 million people to become sick, 128,000 to be hospitalized, and 3,000 to die (CDC 2010)—costing the country $16.3 billion in medical treatment and lost wages. From 1998 to 2016, approximately 300 foodborne outbreaks were reported in the United States. Viral pathogens, predominantly norovirus, caused 59 percent of outbreaks. When the cause of an outbreak could be determined, bacterial pathogens caused 39 percent of the outbreaks.

Twenty percent of the foodborne illnesses were associated with the consumption of fruit or vegetables, predominantly those that are eaten uncooked. These fruit and vegetables have been the vehicle for several high-profile disease outbreaks caused by *Escherichia coli* O157:H7, *Salmonella* spp., *Listeria monocytogenes*, *Cyclospora cayetanensis*, and hepatitis A virus. Among bacterial pathogens, *S. enteritidis* has accounted for the largest number of outbreaks, and *L. monocytogenes* has accounted for the majority of deaths. In addition, multistate outbreaks caused by contaminated fresh produce, such as the 2006 outbreak in baby spinach caused by *E. coli* O157:H7, have remained prominent.

Contamination of produce with human pathogens can occur during production, harvesting, packing, storage, cooling, and transportation. Each step in the food chain affects the safety of fresh and processed refrigerated fruit and vegetables held at or below 4°C (39.2°F) (De Roever 1998).

Microbiology of fresh fruit and vegetables

The growth of pathogenic microorganisms is impacted by pH. Foods with a pH of 4.6 or below are considered high-acid foods, and those with a pH above 4.6 are considered low-acid foods. In general, most fruit have a pH below 4.6 (high-acid food), while the majority of vegetables have a pH above 4.6 (low-acid food). Historically, due to fruit's higher acidity level, fruit have had a much better public health record than vegetables (notable outbreaks have occurred in leafy greens, sprouts, and cucumbers). However, there have been outbreaks associated with low-acid fruit, such as netted melons and papayas. Overall, leafy greens (lettuce and spinach) are the commodities most commonly linked to foodborne diseases of nonanimal origin (Castro-Ibáñez et al. 2017; Da Silva Felício et al. 2015).

Most of the foodborne human pathogens are enteric organisms, which originate in the intestinal tracts of animals, including humans. Enteric organisms are spread by the fecal-oral route, meaning they are shed in the feces of infected animals, including humans, and then ingested by a susceptible host, where they propagate. Examples of enteric organisms include *E. coli* O157:H7, *Salmonella* spp., *C. cayetanesis*, norovirus, and hepatitis A virus. A few pathogenic microorganisms, found naturally in the environment, are called environmental pathogens. *L. monocytogenes* is an important environmental pathogen that has led to several outbreaks in produce. *L. monocytogenes* has been associated with plant material and standing water in storage facilities (Buck et al. 2003; Al-Kharousi et al. 2016; Beuchat 2002, 1996; Tango et al. 2018; Söderqvist et al. 2017).

Microorganisms in refrigerated storage

The general factors that influence microbial growth include nutrients, acidity, temperature, oxygen, and water. In general, produce serves as a sufficient source of nutrients and has adequately high moisture content (water activity) to allow most aerobic microorganisms to grow. As mentioned above, most fruit are considered acidic (with pH less than 4.6) and are less supportive of bacterial growth than vegetables, which are considered low-acid foods (with pH greater than 4.6). However, human pathogens such as *E. coli*, *Salmonella* spp., *L. monocytogenes*, parasites, and viruses are acid-tolerant and can survive or grow at pH levels below 4.6.

The use of refrigeration minimizes the growth of microorganisms and is one of the most widely practiced methods of controlling microbial growth in fresh, perishable foods (Gil et al. 2015) (fig. 4.1). However, refrigeration is not sufficient to keep foods safe from all microbial hazards because it does not inactivate or kill human pathogens. When minimally processed and packaged foods are stored in refrigerated conditions, pathogens such as *Salmonella* spp., *Vibrio cholerae*, and *E. coli* O157:H7 are able to survive for extended periods (Palumbo 1986). In addition, some human pathogens can persist and grow in cold-storage temperatures. *L. monocytogenes* thrives in a wet and cool environment (Beuchat 1996; Denis et al. 2016). Other bacteria, like *Salmonella* spp., while unable to reproduce and grow at temperatures below 4°C (39.2°F), can reproduce and grow if safe holding temperatures are not maintained. Ensuring that refrigerated produce is not held between 4°C (40°F) and 60°C (140°F), known as the danger zone, controls the growth of most human pathogens (Olaimat and Holley 2012; Food Safety and Inspection Service 2021).

The recommended storage temperature for most fresh produce items is below the danger zone. However, many tropical and subtropical fruit (papayas and mangoes are examples) and some vegetables (cucumbers are an example) are chilling-sensitive and may be damaged by low temperatures. Chilling-sensitive produce is usually stored at temperatures in the danger zone and requires extra precautions to keep the microbial population low (Lyons 1973).

The temperature of refrigerated fruit and vegetables may vary greatly during harvesting, storage, distribution, and retail, as well as at the consumer's home. Therefore, many opportunities exist for handling foods outside their recommended storage temperature range. Greater time at temperatures above 4°C produces greater chances for the growth of pathogenic organisms.

Consideration must also be given to the type of produce—for example, ready-to-eat leafy greens that have been prewashed using a commercial washing process to remove most contaminants are less likely to cause illness than nonwashed leafy greens. Therefore, at home,

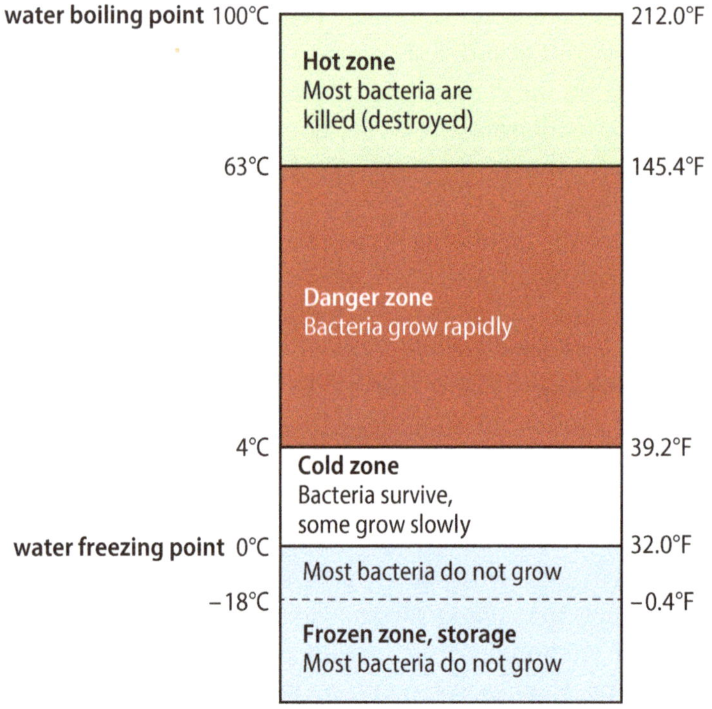

Figure 4.1. Effect of temperature on microorganisms.

nonwashed leafy greens should be thoroughly washed before cutting and eating, especially if consumed raw. A potentially hazardous food is defined by the Food and Drug Administration (FDA) as a food that, to remain safe for human consumption, requires the use of time and temperature control to limit the growth of pathogenic microorganisms or toxin formation. Potentially hazardous foods have been associated with illness in the absence of temperature control, so washing and sanitizing steps need to be taken to keep these foods safe for human consumption. Whole, uncut, raw fruit and vegetables are not classified as potentially hazardous foods. Fresh-cut fruit and vegetables *are* categorized as potentially hazardous foods; these items include leafy greens, salad mixes, and raw seed sprouts (FDA 2013, 642). Regardless, it is imperative that appropriate cooling and storage procedures be used to ensure that fresh produce is safe and retains optimum shelf life—specifically for produce falling in the category of potentially hazardous foods, which must be free of evidence of temperature abuse (Söderqvist et al. 2017).

Regulations to mitigate microbial contamination

In order to safeguard the U.S. food supply, the FDA has established regulations that all growers and food processors must follow, including those that store refrigerated produce. Failure to follow the regulations will cause the FDA to embargo, or remove, produce from the market. If a business does not take action to prevent food safety problems, it may be forced to shut down. While these regulations provide food safety criteria that food businesses must address, each food business has discretion to decide on its actual procedures and actions. The result is the establishment of a food safety management scheme for each business. Such schemes include prerequisite programs (Good Manufacturing Practices, Sanitary Standard Operating Procedures); a food safety plan, including preventive controls formally referred to as critical control points under a management system known as Hazard Analysis and Critical Control Points (HACCP); and a recall plan (FDA 1997).

The Food Safety Modernization Act (FSMA), which went into effect on January 26, 2016, has shifted the focus to industry-wide prevention of foodborne illnesses instead of reaction to individual outbreaks. As a guideline, the FDA has generated seven foundational or major rules to be implemented under the FSMA (fig. 4.2). These rules target different parts of the human and animal food supply chains, and foster food safety by requiring all food–supply chain parties to prevent contamination events (FDA 2015). Some companies, depending on their structure, may be regulated by one, several, all, or additional rules. In the particular case of refrigerated storage facilities, two major rules will typically apply, including the Produce Safety Rule and the Preventive Control for Human Food Rule.

The Produce Safety Rule establishes science-based minimum standards for the safe producing, harvesting, packing, and storing of fruit and vegetables used for human consumption. The rule focuses primarily on controlling biological hazards such as bacteria, viruses, and parasites in the production of fresh produce. It requires food producers to have a food safety program in place and to document adherence to Produce Safety Rule standards. Depending on the type of operation, some refrigerated-storage

Food Safety Modernization Act (FSMA)
21 CFR part 117

7 foundational rules:

- Produce safety rule
- Preventive controls for human food rule
- Intentional adulteration rule
- Foreign supplier verification program (FSVP)
- Accredited third-party certification
- Sanitary transportation rule
- Voluntary qualified importer program (VQIP)

Figure 4.2. Rules for implementing the Food Safety Modernization Act.

facilities for produce may be subject to the requirements of the Produce Safety Rule (FDA 2015; CFSAN 2018).

The Preventive Control for Human Food Rule applies to all facilities that manufacture, process, pack, or hold human food that may be consumed in the United States (FDA 2015). Most of these facilities are required to register with the FDA under section 415 of the Food, Drug, and Cosmetic Act, as amended by FSMA (FDA 2015; CFSAN 2016). Many refrigerated food-storage facilities fall under the Preventive Control for Human Food Rule because they store produce and do not meet the "farm" definition under the FSMA. Some exemptions and modified requirements for the rule may apply, based on a business's size and annual sales (fig. 4.3) (CFSAN 2016). Even if a business has a qualified exemption, it is still required to comply with the portion of the rule related to Current Good Manufacturing Practices, as established by the FDA (CFSAN 2016). The rule establishes standards that cover the basic requirements of processing, packing, and holding food safely. The rule includes requirements for facilities (buildings and equipment), sanitation, human hygiene, processes and controls, pest control, and training, and indicates that food facilities should follow a systematic approach to preventing biological, physical, and chemical hazards from harming food safety (Gil et al. 2015; FDA 1997; CFSAN 2016).

Farms not covered

1. Grow produce identified by FDA as "rarely consumed raw"
2. Grow produce that is not a raw agricultural commodity
3. Grow produce that undergoes a "kill step" through commercial processing
4. Produce grown is only for personal consumption
5. Farms with an average annual value of produce sales <$25,000

Farms with qualified exemptions

Average annual value of produce sales <$500,000
AND
Direct sales to "qualified end users" must be > 50% of all sales

Modified requirements for farms eligible for qualified exemption include

- General provisions (complete business address at point of sale)
- Compliance and enforcement (still remains illegal to sell adulterated food)
- Withdrawal of a qualified exemption (FDA can change status based on risk)

Figure 4.3. Farms with exemptions or modified requirements for the Preventive Control for Human Food Rule.

The FDA has published a document titled "Guidance for Industry: Guide to Minimize Microbial Food Safety Hazards for Fresh Fruits and Vegetables," which addresses microbial food safety hazards and good agricultural and manufacturing practices in the production, harvesting, transportation, washing, sorting, and packing of most fruit and vegetables sold to consumers in an unprocessed or minimally processed form (raw). It describes steps to remove or control field contaminants and can be used by fresh fruit and vegetable producers to help ensure the safety of their produce (CFSAN 2018).

Sanitizers for the food industry are approved either for direct food contact, food surface contact, or use in water. Generally, if a food is considered a raw agricultural commodity [defined by the Code of Federal Regulations (CFR)—Title 21, Chapter 1, Subchapter B, Part 112 (FDA 2021a)], the U.S. Environmental Protection Agency (EPA) retains jurisdiction over the sanitizer. A raw agricultural commodity is a food item in its natural or raw state, including fresh fruit and vegetables that are washed or treated in their unpeeled, natural form, prior to marketing. If the food is considered processed, the FDA regulates the sanitizer. Processed food includes any food other than raw agricultural produce. That is, foods that have been modified from their original state, including raw agricultural commodities that have been subject to cutting, mixing, or packing (fresh-cut fruit and vegetables). If the food is certified as organic, the U.S. Department of Agriculture also becomes involved because operations must additionally comply with the stipulations of the National Organic Program (Agricultural Marketing Service 2021). Disinfectant, sanitizers, and pesticide labels change as new products are evaluated. Updated labels may be retrieved by entering the registration number in the EPA's Pesticide Product Label System.

Facility design and operation

Refrigeration equipment in cold storage should be designed and installed so that it drains without the development of standing water and so that it can be thoroughly cleaned and sanitized to remove microorganisms. Evaporator drip pans may be a source of *L. monocytogenes* and can support the growth of other pathogens, such as Shiga toxin–producing *E. coli* and *Salmonella* spp.

Liquid water forms on evaporator coils after they are defrosted—and in cold-storage rooms for chilling-sensitive commodities, the coils may always be coated with water. The face velocity or the air speed leaving the evaporators should be low enough that water droplets are not carried into the storage space. The airflow system should be designed so that air flows past all interior surfaces, allowing them to dry after cleaning and sanitizing. Equipment surfaces should be kept clean and free of dirt, vegetative materials, and biofilms.

Surfaces in the refrigerated space, such as floors and walls, are a potential source for microbial cross contamination. They should be properly insulated, smooth and made of materials that withstand commonly used cleaners and sanitizers, such as fiberglass-reinforced plastic panels. Equipment in the facility should be designed so that produce residue does not accumulate during operation; and, as mentioned above, so that all parts can be easily accessed for cleaning and so that equipment drains fully, leaving no standing water. This is a particular concern for hydrocoolers, water-spray vacuum coolers, and liquid-ice coolers that have built-in water reservoirs. Cold refrigeration piping in the refrigerated space must be insulated to prevent condensation. The insulation should be covered with a water-impermeable material that can be washed.

Floor-drain piping installed under a concrete floor can be difficult to clean. In refrigerated rooms requiring water drainage, removable, grate-covered gutters allow easy access for periodic cleaning.

Temperature is the key to reducing microbial growth in fresh fruit and vegetables. For cold rooms, an air temperature log or monitoring record is a required part of a food safety plan and is verification that produce is held at a safe temperature. Produce temperature must also be periodically measured to ensure that its temperature is close to the air temperature in the refrigerated space. The facility design and management recommendations described in chapter 2, which covers refrigerated storage, ensure that produce temperature is close to the storage room's air temperature.

Routine documented cleaning and sanitation are required as part of Current Good Manufacturing Practices and food safety plans. In addition, routine and preventive maintenance of refrigeration equipment is essential to maintaining the equipment's proper operation (Gil et al. 2015; FDA 1997).

Chemical (wash-water disinfection) and sanitizing treatments

Raw fruit and vegetables must be thoroughly washed in water to remove soil and other contaminants before they are cut, combined with other ingredients, cooked, served, or offered in ready-to-eat form—with the exception of whole, raw fruit and vegetables intended to be washed by consumers. Water used to rinse, wash, and cool fresh produce must contain 0 colony-forming units per 100 milliliters of generic *E. coli* (CFSAN 2018). This is the same microbial standard that applies to drinking water (Tiemann 2017). Sanitizers are often used to control the microbial content of water used in food processing and storage operations. The Produce Safety Rule and the Preventive Care for Human Food Rule may require environmental monitoring to ensure that pathogens are not present in refrigerated storages. Therefore, to ensure that environmental pathogens such as *L. monocytogenes* do not become resident in the facility, an appropriate sanitizer must be used both for surfaces that contact food and surfaces that do not contact food. Washing fresh produce and applying a sanitizer generally do not reduce microbial populations greater than 2-logs (99%). This reduction in microbial load may enhance shelf life slightly, but it does not ensure food safety, as pathogens may occur at reductions even greater than 2-logs.

Sanitizers, including chemicals for washing, treatment, storage, and processing of fruit and vegetables, shall meet requirements in CFR Title 21, Part 173.315 (FDA 2021b). In addition, manufacturers' application specifications should be followed because the efficacy of sanitizers can be negatively affected by temperature, the presence of biofilms, high organic loads in wash water, and changes in pH.

References

Agricultural Marketing Service. 2021. National Organic Program. https://www.ams.usda.gov/about-ams/programs-offices/national-organic-program

Al-Kharousi, Z. S., N. Guizani, A. M. Al-Sadi, I. M. Al-Bulushi, and B. Shaharoona. 2016. Hiding in fresh fruits and vegetables: Opportunistic pathogens may cross geographical barriers. International Journal of Microbiology 2016:4292417. https://doi.org/10.1155/2016/4292417

Beuchat, L. R. 1996. Pathogenic microorganisms associated with fresh produce. Journal of Food Protection 59(2):204–216. https://doi.org/10.4315/0362-028X-59.2.204

———. 2002. Ecological factors influencing survival and growth of human pathogens on raw fruits and vegetables. Microbes and infection 4(4):413–423. https://doi.org/10.1016/s1286-4579(02)01555-1

Buck, J. W., R. R. Walcott, and L. R. Beuchat. 2003. Recent trends in microbiological safety of fruits and vegetables. Plant Health Progress 4(1):25. https://doi.org/10.1094/PHP-2003-0121-01-RV

Castro-Ibáñez, I., M. I. Gil, and A. Allende. 2017. Ready-to-eat vegetables: Current problems and potential solutions to reduce microbial risk in the production chain. LWT—Food Science and Technology 85(B):284–292. https://doi.org/10.1016/j.lwt.2016.11.073

(CDC) Centers for Disease Control and Prevention. 2010. CDC reports 1 in 6 get sick from foodborne illnesses each year. https://www.cdc.gov/media/pressrel/2010/r101215.html

(CFSAN) Center for Food Safety and Applied Nutrition. 1998. Guidance for industry: Guide to minimize microbial food safety hazards for fresh fruits and vegetables. https://www.fda.gov/regulatory-information/search-fda-guidance-documents/guidance-industry-guide-minimize-microbial-food-safety-hazards-fresh-fruits-and-vegetables

———. 2016. Draft guidance for industry: Hazard analysis and risk-based preventive controls for human food. https://www.fda.gov/Food/GuidanceRegulation/GuidanceDocumentsRegulatoryInformation/ucm517412.htm

———. 2018. Guidance document: Standards for the growing, harvesting, packing, and holding of produce for human consumption. https://www.fda.gov/regulatory-information/search-fda-guidance-documents/draft-guidance-industry-standards-growing-harvesting-packing-and-holding-produce-human-consumption

Da Silva Felício, M. T., T. Hald, E. Liebana, A. Allende, M. Hugas, C. Nguyen-The, G. Skoien Johannessen, T. Niskanen, M. Uyttendaele, and J. McLauchlin. 2015. Risk ranking of pathogens in ready-to-eat unprocessed foods of non-animal origin (FoNAO) in the EU: Initial evaluation using outbreak data (2007–2011). International Journal of Food Microbiology 195:9–19. https://doi.org/10.1016/j.ijfoodmicro.2014.11.005

Denis, N., H. Zhang, A. Leroux, R. Trudel, and H. Bietlot. 2016. Prevalence and trends of bacterial contamination in fresh fruits and vegetables sold at retail in Canada. Food Control 67:225–234. https://doi.org/10.1016/j.foodcont.2016.02.047

De Roever, C. 1998. Microbiological safety evaluations and recommendations on fresh produce. Food Control 9(6):321–347. https://www.sciencedirect.com/science/article/abs/pii/S095671359800022X

(FDA) U.S. Food and Drug Administration. 1997. HACCP Principles and Application Guidelines—National Advisory Committee on Microbiological Criteria for Foods. https://www.fda.gov/food/guidanceregulation/haccp/ucm2006801.htm

———. 2013. Food Code 2013.

———. 2015. Food Safety Modernization Act. 21 C.F.R. 117. 111–353.

———. 2021a. Code of Federal Regulations (CFR) Title 21—Food and Drugs, Chapter I—Food and Drug Administration Department of Health and Human Services, Subchapter B—Food for HUMAN CONSUMPTION, Part 112—Standards for the growing, harvesting, packing, and holding of produce for human consumption. https://www.accessdata.fda.gov/scripts/cdrh/cfdocs/cfcfr/CFRSearch.cfm?CFRPart=112&showFR=1&subpartNode=21:2.0.1.1.12.1

———. 2021b. Code of Federal Regulations (CFR) Title 21—Food and Drugs, Chapter I—Food and Drug Administration, Department of Health and Human Services, Subchapter B—Food for human consumption, Part 173—Secondary direct food additives permitted in food for human consumption, Subpart D—Specific usage additives, Sec. 173.316—Chemicals used in washing or to assist in the peeling of fruits and vegetables. https://www.accessdata.fda.gov/scripts/cdrh/cfdocs/cfcfr/cfrsearch.cfm?fr=173.315

Food Safety and Inspection Service. 2021. "Danger Zone" (40°F–140°F). https://www.fsis.usda.gov/food-safety/safe-food-handling-and-preparation/food-safety-basics/danger-zone-40f-140f#:~:text=Leaving%20food%20out%20too%20long,levels%20that%20can%20cause%20illness

Gil, M. I., M. V. Selma, T. Suslow, L. Jacxsens, M. Uyttendaele, and A. Allende. 2015. Pre- and postharvest preventive measures and intervention strategies to control microbial food safety hazards of fresh leafy vegetables. Critical Reviews in Food Science and Nutrition 55(4):453–68. https://doi.org/10.1080/10408398.2012.657808

Lyons, J. M. 1973. Chilling injury in plants. Annual Review of Plant Physiology 24:445–466. https://doi.org/10.1146/annurev.pp.24.060173.002305

Olaimat, A. N., and R. A. Holley. 2012. Factors influencing the microbial safety of fresh produce: A review. Food Microbiology 32(1):1–19. https://doi.org/10.1016/j.fm.2012.04.016

Palumbo, S. A. 1986. Is refrigeration enough to restrain foodborne pathogens? Journal of Food Protection 49(12):1003–1009. https://doi.org/10.4315/0362-028X-49.12.1003

Scharff, R. L. 2012. Economic burden from health losses due to foodborne illness in the United States. Journal of Food Protection 75(1):123–131.

Söderqvist, K., O. A. Asman, C. Wolff, S. Bertilsson, I. Vågsholm, and S. Boqvist. 2017. Emerging microbiota during cold storage and temperature abuse of ready-to-eat salad. Infection Ecology and Epidemiology 7(1):1328963. https://doi.org/10.1080/20008686.2017.1328963

Tango, C. N., S. Wei, I. Khan, M. S. Hussain, P.-F. N. Kounkeu, J.-H. Park, S.-H. Kim, and D. H. Oh. 2018. Microbiological quality and safety of fresh fruits and vegetables at retail levels in Korea. Journal of Food Science 83(2):386–392. https://doi.org/10.1111/1750-3841.13992

Tiemann, M. 2017. Safe Drinking Water Act (SDWA): A summary of the Act and its major requirements. Congressional Research Service. https://aquadoc.typepad.com/files/crs_sdwa_1march2017.pdf

Chapter 5
Psychrometrics and perishable commodities

James F. Thompson

Psychrometrics is the study of the physical and thermodynamic properties of mixtures of water vapor and air. Commonly used psychrometric variables are dry-bulb temperature, wet-bulb temperature, relative humidity, and dew-point temperature. These terms may be familiar, but often they are not well understood—even though they are important in understanding the interplay between temperature and relative humidity and this interplay's effect on the post-harvest life of perishable commodities.

Psychrometric chart

The psychrometric chart depicts the relationships among the heat and water-vapor properties of air. Figures 5.1 and 5.2 are psychrometric charts that use customary and SI (metric) units, respectively. They illustrate the meaning of various terms.

Air temperature is the horizontal axis of the psychrometric chart. It is measured with a dry-bulb thermometer. The right vertical axis is the moisture content of the air, called the humidity ratio (also called the mixing ratio or absolute humidity). The units of the humidity ratio are mass of water vapor per mass of dry air. Under typical California conditions, the humidity ratio

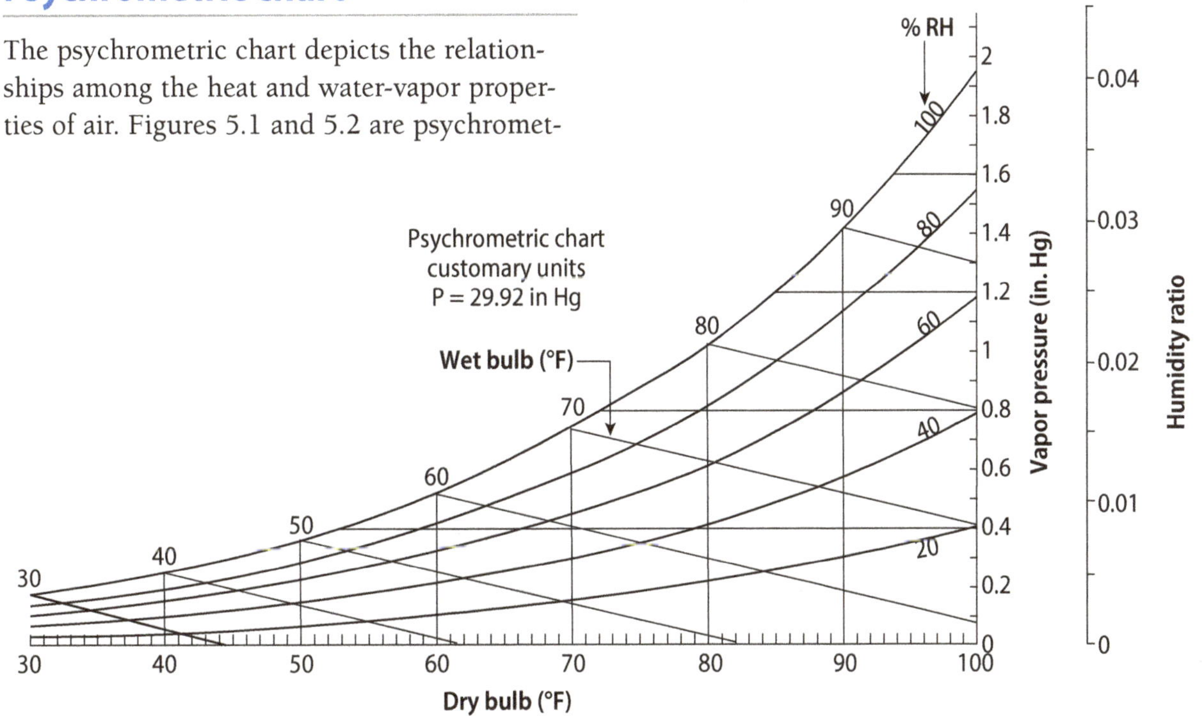

Figure 5.1. A psychrometric chart in customary units for sea-level elevation.

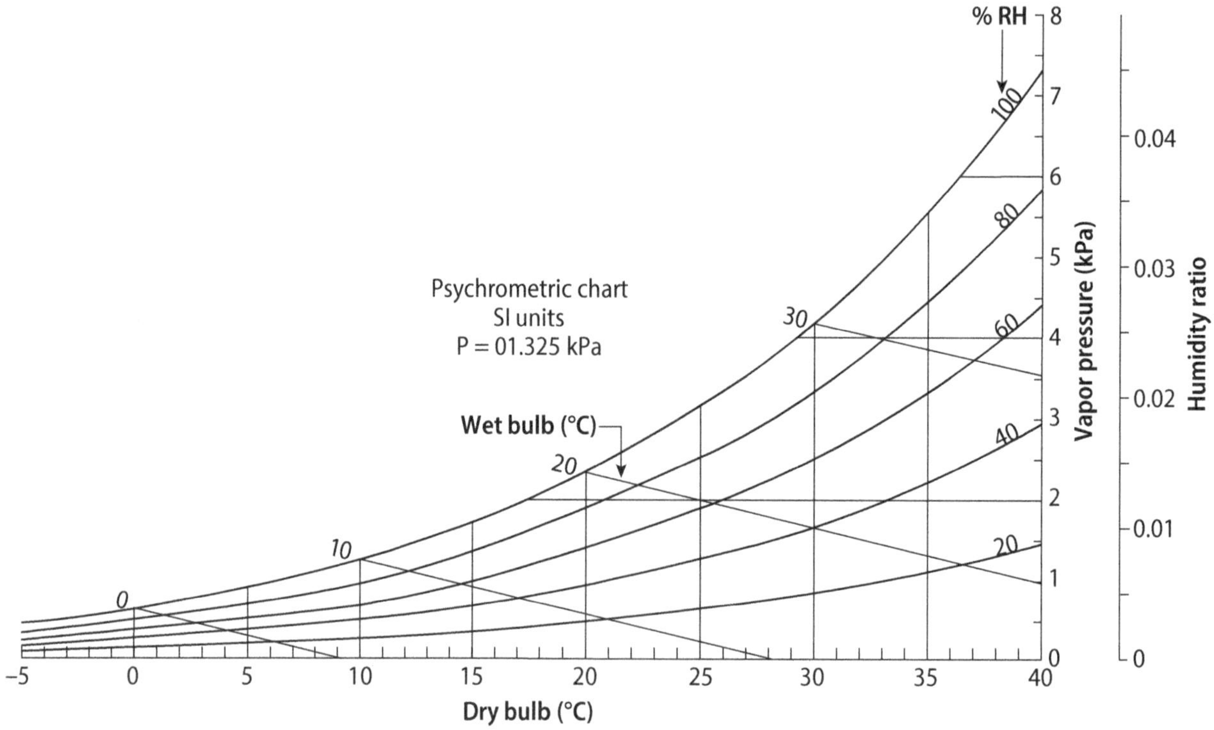

Figure 5.2. A psychrometric chart in SI (metric) units for sea-level elevation.

of outside air varies between 0.004 and 0.015 kilograms per kilogram. Even though water vapor represents only 0.4 to 1.5 percent of the weight of the air, it plays a very significant role in the postharvest life of perishable commodities.

The maximum amount of water vapor that air can hold at a specific temperature is given by the leftmost, upward-curved line in the chart, called the saturation line. Notice that air holds more water vapor at increasing temperatures. As a rule of thumb, the maximum amount of water that the air can hold doubles with every increase of 11°C (20°F) in temperature. The leftmost, upward-curved line is also called the 100 percent relative humidity line.

The 0 percent relative humidity line is the horizontal axis on the chart and represents air with no water vapor in it. The 50 percent relative humidity line is located halfway between the saturation line and the 0 percent relative humidity line. Other relative humidity lines, formed in a similar manner, form a set of lines that curve upward and to the right.

Notice that relative humidity without some other psychrometric variable does not determine a specific air condition on the chart and is not very meaningful. For instance, air at 80 percent relative humidity and 0°C (32°F) has very different properties from air at 80 percent relative humidity and 20°C (68°F).

If a mass of air at a given relative humidity is cooled, but its moisture content doesn't change (that is, its position on the chart moves horizontally left), it loses capacity to hold moisture and its relative humidity increases. If cooled enough, it becomes saturated and intersects the saturation line (100% relative humidity). If cooled further, its position on the chart follows the saturation line downward, and the mass of air loses water in the form of dew or frost. The temperature at which condensation begins to form is called the dew-point temperature if it is above 0°C (32°F) or the frost-point temperature if it is below 0°C (32°F).

Another commonly used psychrometric variable is the wet-bulb temperature. This is represented on the chart by straight lines that slope diagonally downward from left to right. These lines represent the temperature measured by a thermometer covered with water-soaked gauze.

In practice, dry- and wet-bulb temperatures are used to determine an exact point on the psychrometric chart from which the other psychrometric variables can be determined. A psychrometer is an instrument used to measure these variables.

Water vapor pressure is an important property of air in handling perishables. It is directly proportional to the humidity ratio. The following formula can be used to calculate vapor pressure from the humidity ratio:

$$Vp = (w \cdot P_a)/e$$

where

Vp = vapor pressure (same units as P_a; e.g., atmospheres, bars, kP_a, mm Hg, etc.)

w = humidity ratio

P_a = atmospheric pressure

e = 0.622 = ratio of the molecular weight of water to the molecular weight of air.

Psychrometric charts and calculators are based on a specific atmospheric pressure, usually at sea level. Precise calculations of psychrometric variables require an adjustment for barometric pressures that are different from those listed on a standard chart. Consult ASHRAE (2021) for more information. Most field measurements do not require adjustment for pressure.

Effect of psychrometric variables

Each of the psychrometric variables has an effect on produce quality.

Air temperature

Air temperature is the most important variable because it tends to indicate the flesh temperature of perishable commodities. All perishables have an optimal range of storage temperatures. Above the optimum, they respire at unacceptably high rates and lose postharvest life too rapidly. They are also more susceptible to damage from ethylene and disease. In fact, horticultural commodities respire 2.7 times faster for every 10°C (18°F) increase in temperature. Because temperature has such a pronounced effect on storage life, the optimal storage temperature is often just slightly above the lowest temperature that could cause injury. Temperatures below the optimum can result in freezing or chilling damage. Accurate control of temperature is vital in maintaining maximum shelf life.

Vapor pressure

The rate of moisture loss from a perishable is primarily controlled by the difference between the water-vapor pressure in the air in the intercellular spaces of plant material and the water-vapor pressure in the surrounding air. The air in plant material is nearly saturated—in other words, its relative humidity is close to 100 percent. The saturation vapor pressure of this air is therefore determined solely by the temperature of the plant material. From the psychrometric chart it is apparent that low temperatures result in low internal vapor pressures and high temperatures cause high internal vapor pressure.

Consider several examples of how vapor pressure differences affect water loss from perishables (table 5.1 lists water-vapor pressures for these examples). An apple cooled to 0°C (32°F) and placed in a refrigerated room with saturated air (100% relative humidity) at 0°C (32°F) would not lose moisture because the water-vapor pressures of the air in the apple and of the surrounding air would be the same (0.61 kPa). (Actually, the apple would slowly gain moisture because the

Table 5.1. Water-vapor pressure of various storage air conditions and product temperatures

Variables	Water-vapor pressure (kPa)	Water-vapor pressure (lb/in²)
Room air at 0°C (32°F), 100% relative humidity	0.61	0.088
Room air at 0°C (32°F), 70% relative humidity	0.43	0.062
Fresh product* at 0°C (32°F)	0.61	0.088
Fresh product* at 20°C (68°F)	2.34	0.34

Assumes air in product is saturated.

soluble solids in the liquid matrix of the fruit cause the vapor pressure inside the fruit to be slightly lower than if the apple were made of pure water.) However, if the apple were at 20°C (68°F) when it was put into the cold room, the vapor pressure in the apple (2.34 kPa) would be higher than the refrigerated air, and the fruit would lose moisture. If the apple were cooled to 0°C (32°F) but the refrigerated air were at 70 percent relative humidity (0.43 kPa), the apple would still lose water because the vapor pressure in the apple would be higher than in the surrounding refrigerated air. Since the rate of water loss is directly related to the difference in vapor pressure between the apple and the surrounding air, the rate of water loss would be almost ten times greater [9.6 = (2.34 – 0.61)/(0.61 – 0.43)] if the warm apple were stored in air with 100 percent relative humidity than if the cool apple were stored in air with 70 percent relative humidity.

These examples show that moisture loss in storage is minimized by keeping the temperature of the perishable as close as possible to the temperature of the storage air. Also, the air in the storage should be close to 95 percent relative humidity.

Relative humidity

Relative humidity is a term commonly used to describe the humidity of air, but it is not particularly meaningful unless the dry-bulb temperature of the air is known. Together these two variables allow the water vapor pressure to be determined, and vapor pressure is a better index of the potential for water loss. Decay and fungal growth are correlated with relative humidity. At room temperature, no decay organisms can grow below about 65 to 70 percent relative humidity. Fungal growth becomes more prevalent as the relative humidity increases above this level, while bacterial growth is favored above about 95 percent relative humidity.

Dew-point temperature

Condensation of liquid water on the surface of perishables can contribute to disease problems. If a commodity is cooled to a temperature below the dew-point temperature of the outside air and is taken out of the cold room, condensation will form on the surface of the commodity. Condensation can also occur in storage if the air temperature fluctuates too much.

Sample psychrometric calculations

1. A wet-bulb thermometer reads 18°C and a dry-bulb thermometer reads 25°C. What is the relative humidity?

   Solution: In figure 5.3, the diagonal 18°C wet-bulb line and the vertical 25°C dry-bulb line intersect at point A. Point A falls on the 50 percent relative humidity line. The relative humidity is 50 percent.

2. What is the dew-point temperature of the air in problem 1?

   Solution: If the air represented by point A is cooled without changing its moisture content, it will follow a horizontal line until it reaches 14°C at point B. At that temperature, it has 100 percent relative humidity, and any further cooling will cause water to condense out of the air (that is, dew forms). The dew-point temperature is 14°C.

3. What are the humidity ratio and the water-vapor pressure of the air in problem 1?

   Solution: Find the humidity ratio and water-vapor pressure of the air represented by point A by reading horizontally across to the vertical axis of the chart. The humidity ratio (W in fig. 5.3) is 0.01 kilograms per kilogram and the vapor pressure is 1.6 kilopascals.

4. If the air in problem 1 is passed through a 100 percent–efficient evaporative cooler, what will its temperature and relative humidity be after it leaves the cooler?

   Solution: Evaporative cooling (and water-spray humidification) follow the diagonal wet-bulb lines. As air passes through the cooler, it will move from point A along the 18°C wet-bulb line until it reaches 100 percent relative humidity at point C. At this humidity, it is saturated, will not accept any more water, and will stop cooling. It will leave the cooler at 18°C and 100 percent relative humidity.

5. When air represented by point A (dry-bulb temperature = 25°C and wet-bulb temperature = 18°C) enters a storage room with a temperature of 0°C and a relative humidity of 100 percent, will it add moisture to the storage room or dry it out?

   Solution: The air has a dew-point temperature of 14°C (see problem 2). When this air has cooled to just below 14°C it will begin to lose water and continue to lose water until it reaches the storage-room air temperature, point D. In fact, each kilogram of air will lose 0.0006 kilograms of water as it cools. The air will add moisture to the storage room.

6. If air leaves an evaporator coil that is covered with liquid water and it has a temperature of 0°C and 100 percent relative humidity, and it heats 2°C before it reaches stored produce, what is the relative humidity of the air that the produce is exposed to?

   Solution: Sensible heating processes (heating without a change in humidity ratio of the air) follow horizontal lines on the psychrometric chart. Air leaves the evaporator coil at point D and moves horizontally to the right on the chart until it reaches 2°C. At that point the relative humidity will be 83 percent.

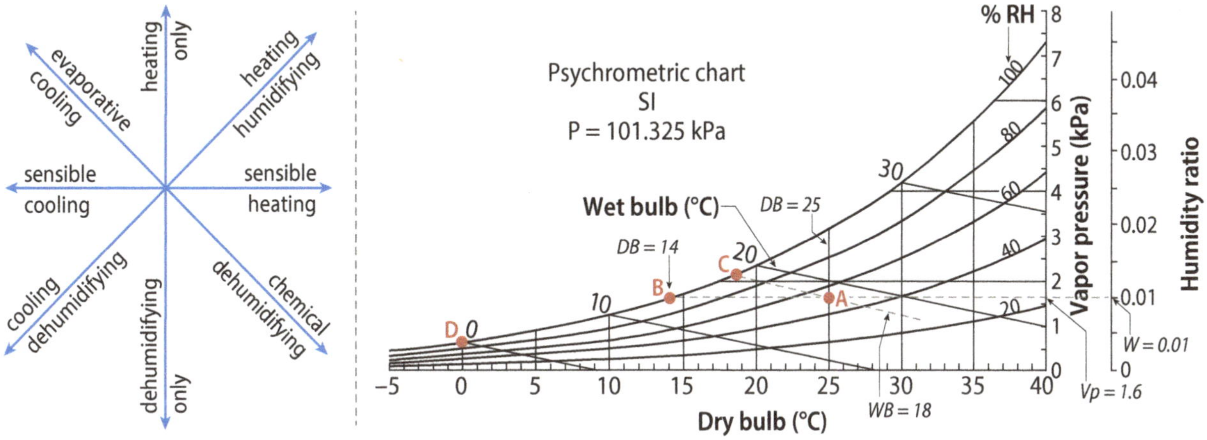

Figure 5.3. Sample psychrometric problems.

Measuring psychrometric variables

All psychrometric properties of air are determined by measuring two psychrometric variables (three, if barometric pressure is considered). For example, if wet- and dry-bulb temperatures are measured, then relative humidity, vapor pressure, dew-point temperature, and so on can be determined with the aid of a psychrometric chart, or a web-based psychrometric calculator. Many variables can be measured to determine the psychrometric state of air, but dry-bulb temperature, wet-bulb temperature, dew-point temperature, and relative humidity are the most commonly measured.

Dry-bulb temperature

Dry-bulb temperature can be simply and inexpensively measured with a glass thermometer. The thermometer should be marked in divisions of 0.2°C, or 0.5°F, if it is to be used in conjunction with a wet-bulb thermometer to determine cold-storage air conditions. The thermometer should be shielded from radiant heat sources such as the sun, motors, lights, external walls, and people. Shielding can be accomplished by placing the thermometer where it cannot "see" warm objects, or by protecting it with a radiant heat-shield assembly.

Hand-held thermistor or thermocouple thermometers can also be used. They are more expensive than glass thermometers but are not necessarily more accurate. Some of these instruments may have a sharp probe, allowing them also to be used to measure produce pulp temperature. Mercury-in-glass thermometers should not be used near food because they are easily broken and any released mercury is toxic. All thermometers must be periodically calibrated against a calibrated thermometer. A stirred ice-water mixture provides an easy way to check calibration at 0°C (32°F).

Wet-bulb temperature

Using a wet-bulb thermometer in conjunction with a dry-bulb thermometer is a common method for determining the state point of the air on the psychrometric chart or a psychrometric calculator. The two thermometers can be in a sling psychrometer that is rotated around a handle or in a battery-operated psychrometer with a fan that draws outside air past the thermometers (fig. 5.4). The wet-bulb thermometer is an ordinary glass thermometer (although electronic temperature-sensing elements can also be used) with a cotton wick secured around the bulb. The wick is wetted with distilled water. Air is forced over the wick, causing it to cool to the wet-bulb temperature. The wet- and dry-bulb temperatures together determine the state point on the psychrometric chart, allowing all other variables to be determined.

An accurate wet-bulb temperature reading depends on the following: the sensitivity and accuracy of the thermometer, maintaining an adequate air speed past the wick, shielding the thermometer from radiation, using distilled or deionized water to wet the wick, and using a cotton wick.

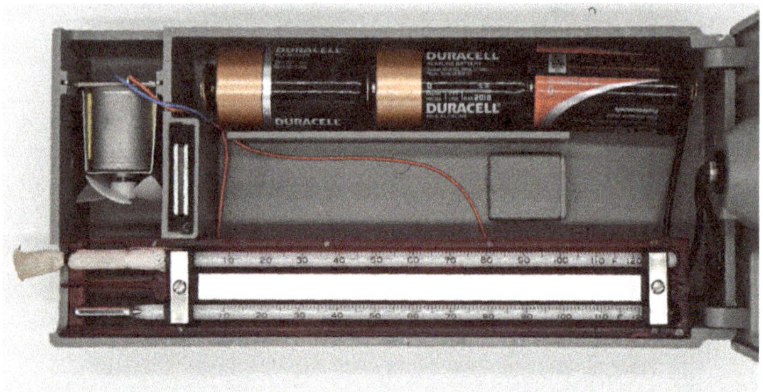

Figure 5.4. Battery-powered psychrometer. *Photo:* Irwin R. Donis-González.

The sensitivity required for a thermometer to accurately determine humidity varies according to the temperature range of the air. More sensitivity is needed at low temperatures than at high ones. For example, at 65°C, a 0.5°C error in the wet-bulb temperature reading results in a 2.6 percent error in relative humidity determination—but at 0°C, that same error results in a 10.5 percent error in relative humidity. In most cases, absolute calibration of the wet- and dry-bulb thermometers is not as important as ensuring that they read the same at a given temperature when the wet-bulb thermometer is dry. For example, if both thermometers measure 0.5°C lower than they should, the resulting error for relative humidity will be less than 1.3 percent at dry-bulb temperatures between 65° and 0°C (and assuming a 5°C difference between dry- and wet-bulb temperatures). Before wetting the wick of the wet-bulb thermometer, operate both thermometers long enough to determine if there is any difference between their readings. If there is a difference, assume that one is correct and adjust the reading of the other accordingly when determining relative humidity.

The rate of evaporation from the wick is a function of the speed of the air passing it. A minimum air speed of about 3 meters per second (500 ft/min) is required for accurate readings. An air speed much below this will result in an erroneously high wet-bulb reading. Wet-bulb devices that do not provide a guaranteed airflow cannot be relied on to give an accurate reading.

As with the dry-bulb thermometer, sources of radiant heat such as motors and lights can affect the wet-bulb thermometer. The reading must be taken in an area protected from these sources of radiation, or thermometers must be shielded from radiant energy.

A build-up of salts from impure water or contaminants in the air affects the rate of water evaporation from the wick and results in erroneous data. Distilled or deionized water should be used to moisten the cotton wick. The wick should be replaced if there is any sign of contamination. The wick material should not have been treated with chemicals such as sizing compounds that affect the water evaporation rate.

Special care must be taken when using a wet-bulb thermometer if the wet-bulb temperature is near freezing. Most humidity tables and calculators are based on a frozen wick at temperatures below 0°C (32°F). At temperatures near or below 0°C, touch the wick with a piece of clean ice or another cold object to induce freezing (distilled water can be cooled below 0°C without freezing). For this method to be accurate, the psychrometric chart or calculator must use frost-bulb temperatures—not wet-bulb temperatures—below 0°C.

Under most conditions, wet-bulb temperature data are not reliable when the relative humidity is below 20 percent or the wet-bulb temperature is above 100°C (212°F). At low humidities, the wet-bulb temperature is much lower than the dry-bulb temperature, and it is difficult for the wet-bulb thermometer to be cooled completely because of heat transferred by the glass or metal stem. At sea level, water boils above 100°C (212°F), so temperatures higher than that cannot be measured with a wet-bulb thermometer.

In general, properly designed and operated wet- and dry-bulb psychrometers can measure relative humidity with an error of less than 2 percent of the actual relative humidity. Improper operation greatly increases the error.

Relative humidity

Hygrometers are used to directly measure relative humidity. Hygrometers often sense relative humidity with an electric sensing element or a mechanical system. Electric hygrometers are based on substances whose electrical properties change as a function of their moisture content (fig. 5.5). As the humidity of the air around the sensor increases, the sensor's moisture increases, proportionally affecting its electrical properties. These devices are more expensive than wet- and dry-bulb psychrometers, but their accuracy is not as severely affected by incorrect operation. An error of less than 2 percent of the actual

Figure 5.5. Sensing element for an electronic relative-humidity sensor. *Photo:* Irwin R. Donis-González.

humidity is often obtainable. Sensors lose their calibration if allowed to become contaminated, and some lose calibration if water condenses on them. Most sensors have a limited life. Calibrate sensors with commercially available kits that allow comparison with the known humidities of saturated-salt solutions.

Mechanical hygrometers usually employ human hairs as a sensing element. Hair changes in length in proportion to the humidity of the air. The response to changes in relative humidity is slow and is not dependable at very high relative humidities. These devices are acceptable as an indicator of a general range of humidity but are not suitable when accurate measurements are required.

Dew-point indicators

Two types of dew-point sensors are commonly used: a saturated-salt system and a condensation dew-point method. The saturated-salt system operates at dew points between –12° and 37°C (10–100°F) with an error of less than ± 1°C (2°F). The saturated-salt system costs less than the condensation system, is not significantly affected by contaminating ions, and has a response time of about 4 minutes. The condensation-type sensor is very accurate over a wide range of dew-point temperatures (less than ± 0.5°C error from –73° to 100°C, or less than ± 1°F error from –100° to 212°F). Condensation dew-point hygrometers can be expensive.

There are a variety of other methods for measuring psychrometric variables. Some are extremely accurate and have characteristics suitable to particular sampling conditions. Most, however, are not commercially available and are used primarily as laboratory instruments.

Reference

(ASHRAE) American Society of Heating, Refrigerating, and Air-Conditioning Engineers. 2021. ASHRAE handbook—fundamentals. Atlanta: American Society of Heating, Refrigerating, and Air-Conditioning Engineers.

Further reading

Gaffney, J. J. 1978. Humidity: Basic principles and measurement techniques. HortScience 13(5):551–555.

Henry, Z. A., G. C. Zoerb, and G. S. Birth. 1991. Instrumentation and measurement for environmental sciences. 3rd ed. St. Joseph, MI: American Society of Agricultural Engineers.

Wexler, A., and W. G. Brombacher. 1951. Methods of measuring humidity and testing hygrometers. Washington, DC: National Bureau of Standards.

Appendices

Appendix A: Storage recommendations for produce, cut flowers, and greens

Appendices A-1 and A-2 summarize postharvest handling environmental information for many products based on the sources listed after the appendices; authors' unpublished data; and industry practice. Recommended conditions and expected storage life for a specific lot of fruit or vegetables may vary from those listed because of cultivar differences, maturity at harvest, growing conditions, grade and quality at harvest, temperature management practices after harvest, and use of special treatments.

Ideal holding temperatures. The temperatures shown in the appendices are generally considered optimal for the commodity. Unless produce is subject to chilling injury, it should be held as close to its freezing temperature as possible. Many commodities respond best if held at 0°C (32°F) or below. Storage near the freezing point is possible only if the cooling system's temperature fluctuations are not great enough to cause freezing.

Products with ideal holding temperatures greater than 0°C (32°F) may be subject to chilling injury if held below their ideal temperature. Injury can take several forms, including surface pitting, tissue discoloration, increased susceptibility to microorganisms, altered flavor or texture, and loss of ability to ripen. Chilling injury is often found only for some varieties or for produce grown under certain climatic conditions. Chilling injury is a time-temperature function, and thus it may be possible to hold some commodities safely for short periods at temperatures below those generally found to cause injury. Holding-temperature requirements may also vary according to ripeness of the produce.

Relative humidity. A relative humidity of 90 to 95 percent is recommended for most fresh commodities. Commodities can only benefit from relative humidity this high if they are uniformly cooled to near their optimal storage temperature. The storage room can be at lower humidity if commodities are packed in film bags, keeping humidity high around the product.

Highest freezing temperature. The freezing points shown in appendix A-1 are the highest noted for any test of that commodity. Because of limitations of test data, readers should be aware that some varieties may have freezing points higher or lower than those shown for a commodity. The freezing point of a commodity is influenced by level of soluble solids. Fruit or vegetables with unusually low soluble-solids content may freeze at temperatures usually considered safe for them. For example, freezing has occurred at temperatures not generally harmful in Bartlett pears when the fruit has low soluble solids, with damage occurring first among the least-ripe pears in the lot.

Market life of produce. There is a maximum time during which a product can remain marketable under good storage and distribution conditions, with considerable variability resulting from cultivar, growing conditions, quality at harvest, or use of special treatments. To obtain the longest market life, produce must be properly handled before as well as during storage. Buyers should communicate closely with suppliers to determine the suitability of a product lot for their marketing and handling chain.

Appendix A–1: Storage recommendations for produce

Common names	Scientific name	Storage temperature		Relative humidity (%)	Highest freezing temperature		Ethylene production	Ethylene sensitivity	Approximate storage life	Beneficial CA and MAP conditions, storage comments
		°C	°F		°C	°F				
acerola, Barbados cherries	*Malpighia glabra*	0	32	85–90	−1.4	29.4	—	—	6–8 weeks	—
African horned melons, kiwanos	*Cucumis metuilferus*	13–15	55–59	90	—	—	L	M	6 months	—
amaranth, pigweed	*Amaranthus* spp.	0–2	32–36	95–100	—	—	VL	M	10–14 days	—
apples, air-stored	*Malus pumila*	−1–4	30–39	90–95	−1.5	29.3	H–L	H	2–4 months	Cultivar, location, and growing conditions affect temperature and CA recommendations
apples, CA-stored	*Malus pumila*	0–2.5	32–36	90–95	−1.5	29.3	H–L	H	3–12 months	0.7–4.5% O_2 and 0.7–4.5% CO_2
apricots	*Prunus armeniaca*	−0.5–0	31–32	90–95	−1.0	30.0	L	M	1–4 weeks	2–3% O_2 and 2–3% CO_2
arazas	*Eugenia* ssp.	12–13	54–55	90–95	—	—	—	—	2 weeks	—
artichokes, globe	*Cynara scolymus*	0	32	>95	−1.2	29.9	VL	L	2 weeks	—
arugula	See salad greens									
Asian pears, nashis	*Pyrus serotina; P. ussuriensis*	−1–1	30–34	90–95	−1.5	29	H–L	H	3–5 months	1–5% O_2 and <2% CO_2
asparagus: green, white	*Asparagus officinalis*	0–2.5	32–37	95–99	−0.6	31.0	M	M	2–3 weeks	5–10% CO_2; T = 2.5° for storage >10 days
atemoyas	*Annona squamosa x; A. chermola*	10–13	50–55	90–95	—	—	H	H	2 weeks	—
avocados, Hass cultivar	*Persea Americana*	4–7	39–45	85–95	−1.6	29.1	H	H	2–4 weeks	2–5% O_2 and 3–10% CO_2; early season, 5–7°
avocados, nonspecific cultivar and unripe	*Persea Americana, P. nubigena*	5–12	41–54	85–95	—	—	H	H	—	Optimum conditions vary by cultivar
babacos, mountain papayas	*Carica candamarcensis*	7	45	85–90	—	—	—	—	1–2 weeks	—

Continued on next page

Appendix A–1: Storage recommendations for produce, continued

Common names	Scientific name	Storage temperature		Relative humidity (%)	Highest freezing temperature		Ethylene production	Ethylene sensitivity	Approximate storage life	Beneficial CA and MAP conditions, storage comments
		°C	°F		°C	°F				
bananas	*Musa acuminate x; M. balbisiana*	13.3–14.4	56–58	90–95	−0.8	30.5	M	H	1–4 weeks	2–5% O_2 and 2–5% CO_2. Remove ethylene. Use 14.4° for winter shipments, 13.3° for nonwinter
Barbados cherries	See acerola									
beans: fava, broad	*Vica faba*	0	32	90–95	—	—	—	—	1–2 weeks	—
beans, lima	*Phaseolus linatus*	5–6	41–43	95	−0.6	31.0	L	M	5–7 days	—
beans, long	*Vigna sesquipedalis*	5–7.5	41–46	95–100	—	—	VL	H	8–12 days	2–5% O_2 and 3–10% CO_2; remove ethylene
beans, runner, flat	*Phaseolus coccineus*	5–7.5	41–46	95–100	—	—	VL	H	8–12 days	2–5% O_2 and 3–10% CO_2; remove ethylene
beans: snap, yellow wax, green, string	*Phaseolus vulgaris*	5–7.5	41–46	95–100	−0.7	30.7	VL	H	8–12 days	2–5% O_2 and 3–10% CO_2; remove ethylene
beans, winged	*Psophocarps tetragonolo-bus*	10	50	90	—	—	—	—	4 weeks	—
beets, bunched	*Beta vulgaris*	0	32	>98	−0.4	31.3	VL	L	10–14 days	little or no benefit from CA or MAP
beets, topped	*Beta vulgaris*	1–2	34–36	98	−0.9	30.3	VL	L	8–10 months	little or no benefit from CA or MAP
Belgian endive	See chicory									
bitter melons, bitter gourd	*Momordica charantia*	10–12	50–54	85–90	—	—	L	M	2–3 weeks	2–3% O_2 and 5% CO_2
blackberries	*Rubus* sp.	−0.5–0	31–32	>90	−0.8	30.6	M–VL	—	2–14 days	5–10% O_2 and 10–20% CO_2
black salsify, scorzonera	*Scorzonera hispanica*	0–1	32–34	95–98	—	—	VL	L	6 months	—
blueberries	*Vaccinium* sp.	−0.5–0	31–32	>90	−1.3	29.7			2–4 weeks	1–10% O_2 and 10–15% CO_2
bok choy, Chinese chard	*Brassica campestris* ssp. *chinensis*	0–5	32–41	>95	−0.5	31	L	M	3 weeks	3% O_2 and 5% CO_2

Continued on next page

Appendix A-1: Storage recommendations for produce, continued

Common names	Scientific name	Storage temperature		Relative humidity (%)	Highest freezing temperature		Ethylene production	Ethylene sensitivity	Approximate storage life	Beneficial CA and MAP conditions, storage comments
		°C	°F		°C	°F				
breadfruit	*Artocarpus altilis*	12–14	54–57	90–95	—	—	L	M	20 days	2–5% O_2 and 5% CO_2
broccoli	*Brassica oleracea* var. *ilalica*	0	32	98–100	−0.6	31.0	L	H	2–3 weeks	1–2% O_2 and 5–10% CO_2
Brussels sprouts	*Brassica oleracea* var. *gemmifera*	0	32	95–100	−0.8	30.5	L	H	3–5 weeks	1–2% O_2 and 5–10% CO_2
cabbage, early crop	*Brassica oleracea* var. *capatata*	0	32	98–100	−0.9 to −0.83	30.4–30.5	VL	M	3–6 weeks	2–3% O_2 and 4–5% CO_2
cabbage, late crop	*Brassica oleracea* var. *capatata*	0	32	95–100	−0.9 to 0.83	30.4–30.5	VL	M	5–6 months	2–3% O_2 and 4–5% CO_2
cactus fruit	See prickly pears									
carambolas	*Averrhoa carambola*	4–5	39–41	90–95	−1.2	29.8	M	—	3–5 weeks	—
carrots, bunched	*Daucus carata*	0	32	90–100	—	—	VL	H	8–12 days	—
carrots, topped	*Daucus carata*	0	32	90–100	−1.4	29.5	VL	H	5–9 months; immature, 2–3 weeks	High relative humidity extends storage life; no CA benefit
cashew apples	*Anacardium occidentale*	0–2	32-36	85–90	—	—	—	—	5 weeks	—
cassava, yucca, manioc	*Manihot esclenta*	0–5	32–41	85–90	—	—	M	L	1 month	Waxing extends life to >30 days
cauliflower	*Brassica oleracea* var. *botrytis*	0	32	95–98	−0.8	30.6	L	H	3 weeks	2–5% O_2 and 2–5% CO_2
celeriac	*Apium graveolens* var. *rapaceum*	0–2	32-36	97–98	−0.9	30.4	VL	L	6–8 months	Not recommended, although 2% O_2 and 2–3% CO_2 may be beneficial
celery	*Apium greveolens*	0	32	95	−0.5	31	VL	M	5–7 weeks	2–4% O_2 and 3–5% CO_2 provide small to moderate benefit

Continued on next page

Appendix A-1: Storage recommendations for produce, continued

Common names	Scientific name	Storage temperature		Relative humidity (%)	Highest freezing temperature		Ethylene production	Ethylene sensitivity	Approximate storage life	Beneficial CA and MAP conditions, storage comments
		°C	°F		°C	°F				
chard	*Beta vulgaris* var. *cicla*	0	32	95–100	—	—	VL	H	10–14 days	—
chayotes	*Sechium edule*	7	45	85–90	—	—	—	—	4–6 weeks	—
cherimoyas, custard apples	*Annona cherimola*	10–13	50–55	90–95	−2.2	28.0	H	H	2–3 weeks	—
Cherries, sour	*Prunus cerasus*	0	32	90–95	−1.7	29.0	VL	L	3–7 days	3–10% O_2 and 10–12% CO_2
cherries, sweet	*Prunus avium*	−1–0	30–32	>95	—	—	VL	L	2–4 weeks	1–5% O_2 and 5–20% CO_2, MAP beneficial
chicory, Belgian endive, Witloof chicory	*Cichorium intybus*	0	32	95–100	—	—	VL	M	>2–4 weeks	3–4% O_2 and 4–5% CO_2; light causes greening if temperature >0°C
chiles	See peppers									
Chinese artichokes	*Stachys affinis*	0	32	90–95	—	—	VL	VL	1–2 weeks	—
Chinese broccoli, gai lan	*Brassica oleracea* var. *alboglabra*	0	32	95–100	—	—	VL	H	10–14 days	—
Chinese cabbage, Napa	*Brassica rapa* ssp. *pekinensis*	0	32	98	−0.9	30.4	VL	H	3–6 months	1–2% O_2 and 2–6% CO_2
coconuts	*Cocos nucifera*	0–1.5	32–35	75–85	−0.9	31.1	VL	L	2 months	storage time for mature, dehusked
collards, kale	*Brassica oleracea* var. *Acephala*	0	32	95–98	−0.5	31.1	VL	H	3 weeks	CA and MAP not used commercially
corn: sweet, baby	*Zea mays*	0	32	95–98	−0.6	31.0	VL	L	5–8 days	2–4% O_2 and 5–10% CO_2
cranberries	*Vaccinium macrocarpon*	2–5	36–41	90–95	−0.9	30.3	L	L	2–4 months	CA not used commercially
cucumbers	*Cucumis sativus*	10–12.5	50–55	95	−0.5	31.1	L	M	2 weeks	little benefit
currants, black	*Ribes nigrum*	−0.5–0	31–32	95	−1.0	30.2	L	L	1.5 weeks	15–20% CO_2 extends storage to 3 weeks

Continued on next page

Appendix A-1: Storage recommendations for produce, continued

Common names	Scientific name	Storage temperature		Relative humidity (%)	Highest freezing temperature		Ethylene production	Ethylene sensitivity	Approximate storage life	Beneficial CA and MAP conditions, storage comments
		°C	°F		°C	°F				
currants: red, white, pink	*Ribes sativum*	−0.5–0	31–32	95	−1.0	30.2	L	L	2.5 weeks	2% O_2 and 18–20% CO_2 at 1°C extends storage to 8–14 weeks
custard apples	See cherimoyas									
daikon, lobok, Oriental radish	*Raphanus sativus*	0–1	32–34	95–100	—	—	VL	L	4 months	—
dasheen	See taro									
dates	*Phoenix dactylifera*	0	32	70–75	−15.7	3.7	VL	L	1 year; >1 year at −17.5°C	Dry dates at 20°C can be stored for many years
dewberries	*Rubus* spp.	−0.5–0	31–32	90–95	−1.3	29.7	L	L	2–3 days	—
dragon fruit, pitayas	*Hylocereus* spp.	10	50	—	—	—	VL	—	2 weeks	5 weeks in MAP
durian	*Durio zibethinus*	13	55	85–95	—	—	H	—	6–8 weeks	Life varies by cultivar; fully ripe storage temperature = 4.4°C
eggplants	*Solanum melongena*	10–12	50–54	90–95	—	—	M	L	2 weeks	little benefit of CA; MAP beneficial
elderberries	*Sambucus canadensis*	−0.5–0	31–32	95	—	—	—	—	5–14 days	—
endive and escarole	*Chicorium endiva*	0	32	95–100	−0.1	31.7	VL	M	2–3 weeks	
feijoas, pineapples, guavas	*Feijos sellowiana*	5–10	41–50	90	—	—	M	L	2–3 weeks	—
fennel	*Foeniculum vulgare*	0	32	90–95	—	—	L	—	2 weeks	—
figs, fresh	*Ficus carica*	−1–0	30–32	90–95	−2.4	27.6	L	M	1–2 weeks	3–4 weeks in 5–10% O_2 and 15–20% CO_2
garlic	*Allium sativum*	−1–0	30–32	60–70	−0.8	30.5	VL	L	>9 months	5–15% CO_2 retards sprouting and decay
ginger	*Zingiber officinale*	12–14	54–57	85–90	—	—	VL	L	2–3 months	—

Continued on next page

Appendix A–1: Storage recommendations for produce, continued

Common names	Scientific name	Storage temperature		Relative humidity (%)	Highest freezing temperature		Ethylene production	Ethylene sensitivity	Approximate storage life	Beneficial CA and MAP conditions, storage comments
		°C	°F		°C	°F				
ginseng	*Panax* spp.	0	32	≥95	—	—	VL	L	2 months	1% O_2 and >5% CO_2
gooseberries	*Ribes grossularia*	−05–1	31–34	95	−1.1	30.0	L	L	3 weeks	6–8 weeks at 1°C, 1.5% O_2 and 10–15% CO_2
granadillas	See passion fruit									
grapefruit	*Citrus paradisi*	5–8	41–46	95	−1.1	30	VL	M	6 weeks	Wax with fungicide reduces chilling injury and decay
grapes, American	*Vitus labrusca*	−0.5–0	31–32	85–90	−1.4	29.4	VL	L	4–7 weeks	not commercially used
grapes, mucsadine	*Vitus rotundifolia*	−0.5–0	31–32	>95	—	—	VL	—	1–4 weeks	10% O_2 and 10–15% CO_2
grapes, table	*Vitus vinifera*	−1–0	30–32	90–95	−3.0 berries	27 berries	VL	L	1–6 months	SO_2 for decay control
greens: turnip, mus-tard, rape	*Brassica* spp.	0	32	95–98	—	—	VL	H	2–3 weeks	
guavas	*Psidium guajava*	8–10	46–50	90–95	—	—	M	M	2–3 weeks	MAP beneficial
Herbs, annual										
basil	*Ocimum basilicum*	12	54	95–100	—	—	VL	H	2 weeks	MAP beneficial
chervil	*Anthriscus cerefolium*	0	32	90–100	—	—	VL	H	1 week	MAP beneficial
coriander, cilantro	*Coriandrum sativum*	0	32	90–100	—	—	VL	H	2 weeks	MAP beneficial
dill	*Anethum graveolens*	0	32	90–100	−0.7	30.7	VL	H	2 weeks	—
savory	*Satureja montana*	0	32	90–100	—	—	VL	H	3 weeks	—
Herbs, perennial										
Chinese chives	*Allium tuberosum*	0	32	90–100	—		L	H	1–2 weeks	—
chives	*Allium schoenopra-sum*	0	32	95–100	−0.9	30.4	L	H	2–3 weeks	MAP beneficial

Continued on next page

Appendix A-1: Storage recommendations for produce, continued

Common names	Scientific name	Storage temperature		Relative humidity (%)	Highest freezing temperature		Ethylene production	Ethylene sensitivity	Approximate storage life	Beneficial CA and MAP conditions, storage comments
		°C	°F		°C	°F				
epazote, wormseed	*Dysphania ambrosioides*	0–5	32–41	90–95	—		VL	M	1–2 weeks	—
marjoram	*Origanum hortensis*	0	32	90–95	—		L	H	1–2 weeks	—
oregano	*Origanum vulgare*	0	32	90–95	—		L	H	1–2 weeks	—
peppermint	*Mentha perperita*	0	32	95–100	—		L	H	2–3 weeks	—
spearmint	*Mentha spicata*	0	32	95–100	—	—	L	H	2–3 weeks	—
rosemary	*Rosmarinus officinalis*	0	32	90–95	—	—	L	H	2–3 weeks	—
sage	*Salvia officinalis*	0	32	90–95	—	—	L	H	2–3 weeks	—
shiso	*Perilla frutescens* var. *crispa*	10	50	95	—		VL	M	1 week	—
tarragon	*Artemesia dracunculus*	0	32	90–95	—	—	L	H	1–2 weeks	—
thyme	*Thymus vulgaris*	0	32	90–95	—	—	L	H	2–3 weeks	—
horseradish	*Amoracia rusticana*	0	32	98–100	−1.8	28.7	L	L	8–12 months	CA has little benefit
husk tomatoes	See tomatillo									
jaboticaba	*Plinia cauliflora*	13–15	55–59	90–95	—	—			2–3 days	—
jackfruit	*Artocarpus heterophyllus*	13	55	85–90	—	—	M	M	2–4 weeks	—
Jerusalem artichokes	*Helianthus tuberosus*	0–2	32–34	90–95	−2.5	27.5	VL	L	6–12 months	—
jicamas, yam beans	*Pachyrhizus erosus*	12.5–15	55–59	80–90	—	—	VL	L	2–4 months	—
jujubes, Chinese dates	*Ziziphus jujuba*	2.5–10	37–50	85–90	−1.6	29.2	L	M	1 month	—
kakis	See persimmons									

Continued on next page

Appendix A-1: Storage recommendations for produce, continued

Common names	Scientific name	Storage temperature		Relative humidity (%)	Highest freezing temperature		Ethylene production	Ethylene sensitivity	Approximate storage life	Beneficial CA and MAP conditions, storage comments
		°C	°F		°C	°F				
kiwanos	See African horned melons									
kiwifruit	*Actinida deliciosa*	0	32	90–95	−0.9	30.4	VL	H	4–5 months	1–2% O_2 and 3–5% CO_2 extends life to 6 months
kohlrabi	*Brassica oleracea* var. *Gongylodes*	0	32	98–100	−1.0	30.2	VL	L	2–3 months	no CA benefit; perforated film bags reduce water loss
leeks	*Allium porrum*	0	32	95–100	−0.07	30.7	VL	M	2–3 months	1–3% O_2 and 2–5% CO_2 extends life to 4–5 months
lemons	*Citrus limon*	7–12	45–54	85–95	−1.4	29.4	VL	M	1–6 months	CA rarely used; need good storage ventilation
lettuce	*Latuca sativa various* ssp.	0	32	98–100	−0.2	31.6	VL	H	2–4 weeks	3% O_2 and 2% CO_2 for 1-month transport; bin liners or head wraps maintain high RH
limes: Persian, Key	*Citrus latifolia, C. aurantifolia*	10	50	95	−1.6	29.1	VL		8 weeks	very limited use of CA
lobok	See daikon									
loganberries	*Rubus* spp.	−0.5–0	31–32	90–95	−1.7	29.0	L	L	2–3 days	
longans	*Dimocarpus longan*	4–7	39–45	90–95	−2.4	27.7	VL	L	2–3 weeks	SO_2 increases storage life
longkong, langsat, duku	*Lansium domesticum*	8	46	90	—	—	—		3 weeks	5% O_2 and 0% CO_2
loquats	*Eriobotrya japonica*	0–5	32–41	>90	−1.9	28.6	L	L	3–4 weeks	Film bagging retards water and acids loss
luffas, Chinese okra, sponge gourds	*Luffa* spp.	10–12	50–54	90–95	—	—	VL	M	2 weeks	—
lychees	*Litchi chinensis*	2–5	36–41	90–95	—	—	L	M	3–5 weeks	3–5% O_2 and 5% CO_2 at 5–7°C
mamey apples	*Mammea americana*	13–15	55–59	90–95	—	—	VH	H	2–3 weeks	—

Continued on next page

Appendix A–1: Storage recommendations for produce, continued

Common names	Scientific name	Storage temperature		Relative humidity (%)	Highest freezing temperature		Ethylene production	Ethylene sensitivity	Approximate storage life	Beneficial CA and MAP conditions, storage comments
		°C	°F		°C	°F				
mandarins, tangerines	*Citrus reticulata, C. unshiu*	5–8	41–46	95	−1.1	30.1	VL	M	2–4 weeks	—
mangoes	*Mangifera indica*	10–13	50–55	85–90	−1.4	29.5	M	M	2–4 weeks	3–5% O_2 and 5–10% CO_2 at 7–9°C
mangosteens	*Garcinia mangostana*	12–14	54–57	85–90	—	—	H	H	20 days	5% O_2 and 5% CO_2 extend life to 30 days
melons, cantaloupes, Persian, netted	*Cucumis melo* var. *reticulatus*	2–7	36–45	95	−1.2	29.9	H	M	10–14 days	3–5% O_2 and 10–20% CO_2
melons, casabas	*C. melo*	10	50	90–95	−1.0	30.3	L	L	3 weeks	limited commercial use
melons, Crenshaw	*C. melo*	10	50	90–95	−1.1	30.1	M	M	3 weeks	limited commercial use
melons, honeydew, orange flesh	*C. melo*	7	45	95	−1.1	30.1	M	H	7–10 days	limited commercial use
mushrooms	*Agaricus* and other genera	0–1	32–34	95	−0.9	30.4	VL	M	7–9 days	CA rarely used; overwrap packaging slows moisture loss
mustard greens	*Brassica juncea*	0	32	90–98	—	—	VL	H	1–2 weeks	—
nashis	See Asian pears									
nectarines	*Prunus persica* var. *nectarina*	−1–0	30–32	90–95	−3 to −1.5	27–30	VL unripe; H ripe	M	2–4 weeks	1–2% O_2 and 3–5% CO_2
nopalitos, cactus leaves	*Opuntia* spp.	5	41	95–99	—	—	VL	L	3 weeks	—
okra	*Abelmoschus esculentus*	7–10	45–50	>90	−1.8	28.7	L	M	1–2 weeks	CA has slight benefit
olives, fresh green	*Olea e uropaea*	5–7.5	41–46	90–95	−1.4	29.4	VL	M	4–6 weeks	2–3% O_2 and 0–1% CO_2 at 5°C extends life to 12 weeks
onions, green bunched	*Allium cepa*	0	32	95–98	−0.9	30.4	VL	L	3–4 weeks	6–8 weeks storage at 2% O_2 and 5% CO_2

Continued on next page

Appendix A–1: Storage recommendations for produce, continued

Common names	Scientific name	Storage temperature		Relative humidity (%)	Highest freezing temperature		Ethylene production	Ethylene sensitivity	Approximate storage life	Beneficial CA and MAP conditions, storage comments
		°C	°F		°C	°F				
onions, mature dry bulbs	*Allium cepa*	0	32	65–75	−0.8	30.6	VL	L	pungent, 6–9 months; mild, 1–3 months	3% O_2 and 5% CO_2 for pungent onions
oranges, California and Arizona	*Citrus sinensis*	3–8	37–46	90–95	−0.8	30.6	VL	M	<12 weeks	CA does not inhibit decay, rarely used
oranges, Florida and Texas	*Citrus sinensis*	0–1	32–34	85–90	−0.8	30.6	VL	M	<12 weeks	CA does not inhibit decay, rarely used
papayas	*Carica papaya*	7–13	45–55	90–95	−0.9	30.3	M	M	1–2 weeks	36-day life at 3% O_2 and 8% CO_2 and 10°C; not commercially used
parsley	*Petroselinum crispum*	0	32	95–100	−1.1	30.0	L	H	1–2 months	MAP beneficial
parsnips	*Pastinaca sativa*	0–1	32–34	98	−0.9	30.4	VL	H	4–6 months	Ethylene causes bitterness
passion fruit, granadillas	*Passiflora* spp.	7–10	45–50	90–95	—	—	VH	L	2 weeks	—
peaches	*Prunus persicae*	−1–0	30–32	90–95	−3 to −1.5	27–30	VL unripe; H ripe	M	2–4 weeks	6% O_2 and 17% CO_2 at 0°C reduces internal browning
pears, European	*Pyrus communis*	−1	30	90–94	−1.7	29.0	H– VL	M	2–7 months	2–2.5% O_2 and 0.8–1 % CO_2
peas: green, snow, sugar, snap	*Pisum sativum*	0	32	95–98	−0.6	30.9	VL	M	1–2 weeks	—
peppers: bell, chili, pimiento	*Capsicum annul*	7	45	90–95	−0.7	30.7	L	M–L	2–3 weeks	Bagging adds 1 week; CA has slight benefit
persimmons, kakis	*Diospyrus kaki*	−1–1	30–34	90–95	−2.0	28.4	L	H	3 months	5 months in 3–5% O_2 and 5–8% CO_2 and 0% ethylene
pineapples	*Annas comosus*	7–12	45–54	85–95	−1.1	30.0	L	L	14–20 days	1–2% O_2 and 0–10% CO_2 delays chilling injury
plantains	*Musa acuminate x; M. balbisiana*	9–12	48–53	90–95	−0.8	30.5	M	H	1–4 weeks	2–5% O_2 and 2–5% CO_2; remove ethylene

Continued on next page

Appendix A-1: Storage recommendations for produce, continued

Common names	Scientific name	Storage temperature °C	Storage temperature °F	Relative humidity (%)	Highest freezing temperature °C	Highest freezing temperature °F	Ethylene production	Ethylene sensitivity	Approximate storage life	Beneficial CA and MAP conditions, storage comments
plums, fresh prunes	*Prunus salicina*	−1.1–0	30–32	90–95	−2 to −1	28–30	M–VL	M	2–5 weeks	1–2% O_2 and 3–5% CO_2
pomegranates	*Punica granatum*	5–7.2	41–45	90–98	−3.0	26.6	VL	L	2–3 months	3% O_2 and 6% CO_2
potatoes, early, immature	*Solanum tuberosum*	10–15	50–59	90–95	−0.8	30.6	VL	M	10–14 days	minimal CA benefit
potatoes: late crop, fresh market	*Solanum tuberosum*	7–10	45–50	95–98	−0.8	30.6	VL	M	2–12 months	minimal CA benefit
prickly pears	*Opuntia* spp.	5–8	41–46	90–95	—	—	L	L	2–5 weeks	2% O_2 and 2–5% CO_2 at 5°C
pumpkins, winter squash, acorn, butter-nut, crookneck	*Cucurbita* spp.	10–13	50–55	50–70	−0.8	30.6	VL	M	2–6 months	Storage life depends on species/variety; some varieties benefit from CA
quinces	*Cydonia oblonga*	−0.5–0	31–32	90	−2.0	28.4	—	—	2–3 months	—
radicchio	*Chichorium intybus*	3–5	37–41	99–100	—	—	L	M	1–4 weeks	3% O_2 and 5% CO_2; 3-month storage at −2° to −3°C
radishes	*Raphuanus sativus*	0	32	90–95	−0.7	30.7	VL	L	3–4 weeks	1–2% O_2 and 2–3% CO_2; 1–2 week storage with tops
rambutans	*Nephelium lappacium*	8–15	46–59	90–95	—	—	VL	L	14–16 days	3–5% O_2 and 7–12% CO_2 adds 4–5 days of life
raspberries	*Rubus ideaus*	−05–1	31–33	>95	−0.9	30.4	M	L	2–5 days	5–10% O_2 and 10–20% CO_2
rhubarb	*Rheum rhabarbarum*	0	32	95–100	−0.9	30.3	VL	L	2–4 weeks	
rutabaga, swede	*Brassica Napus*	0	32	98–100	−1.1	30.1	VL	L	4–6 months	no CA benefit
salad greens: corn salad, dandelion, sorrel, miner's lettuce, purs-lane, mizuna, arugula	Various species	0–2	32–36	95–100	—	—	VL	H	1–2 weeks	MAP is beneficial in reducing weight loss

Continued on next page

Appendix A–1: Storage recommendations for produce, continued

Common names	Scientific name	Storage temperature		Relative humidity (%)	Highest freezing temperature		Ethylene production	Ethylene sensitivity	Approximate storage life	Beneficial CA and MAP conditions, storage comments
		°C	°F		°C	°F				
salsify, black, scorzonera	*Scorzonera hispanica*	0	32	95–98	−1.1	30.1	VL	L	3–4 months	6 months at 3% O_2 and 3% CO_2
salsify, vegetable oyster	*Tragopogon porrifolius* var. *sativus*	0	32	95–98	−1.1	30.1	VL	L	3–4 months	—
sapodillas, sapotas, chikus, chicos	*Manikara achras*	12–16	54–61	85–90	—	—	M	H	2–3 weeks	MAP and ethylene removal extend storage life
sapotes, black	*Diospyros dignya*	13–15	55–59	85–90	−2.3	27.8	—	—	2–3 weeks	—
sapotes, canistels, yellow	*Pouteria campechi-ana*	13–15	55–59	85–90	−1.8	28.7	—	—	3 weeks	—
sapotes, mamey	*Pouteria sapota*	13–18	55–64	85–90	—	—	VH	—	2–6 weeks	—
sapotes, white	*Casimiroa edulis*	19–21	66–70	85–90	−2.0	28.4	—	—	2–3 weeks	—
scorzonera	See black salsify									
shallots	*A. cepa* var. *ascalonicum*	0–2.5	32–37	65–70	−0.7	30.7	L	L	—	—
soursops	*Annona muricata*	0	32	85–90	—	—	—	—	1–2 weeks	—
Southern peas, cowpeas	*Vigna unguiculata*	4–5	39–41	95	—	—	—	—	6–8 days	—
spinach	*Spinacia oleraceae*	0	32	95–98	−0.3	31.5	VL	H	10–14 days	7–10% O_2 and 5–10% O_2; CA and MAP not commonly used
sprouts from seeds	Many spp.	0	32	95–100	—	—	L–VL		5–10 days	MAP extends storage life
star apples, caimitos, sweetsops	*Chryso-phyllum cainito*	3–6	37–43	90	−1.2	29.9	H	L	3 weeks	Pick half-ripe, cure for 2 days for maximum stor-age life
strawberries	*Fragaria x ananassa*	0	32	90–95	−0.8	30.6	VL	L	7–10 days	MAP with 10–15% CO_2 retards Botrytis cinerea
summer squash, crook-neck, zucchini	*Cucurbita pepo*	5–10	41–50	95	−0.5	31.1	L	M	1–2 weeks	—

Continued on next page

Appendix A–1: Storage recommendations for produce, continued

Common names	Scientific name	Storage temperature °C	Storage temperature °F	Relative humidity (%)	Highest freezing temperature °C	Highest freezing temperature °F	Ethylene production	Ethylene sensitivity	Approximate storage life	Beneficial CA and MAP conditions, storage comments
sweet corn, baby corn	*Zea mays* var. *rugosa*	0	32	95–98	–0.6	31.0	VL	L	1–2 weeks	2–4% O_2 and 5–10% CO_2
sweet potatoes	*Ipomoea batatas*	13–15	55–59	90	–1.3	29.7	VL	M	12 months	CA not used commercially
Swiss chard	*Beta vulgaris* var. *cycla*	0	32	95–98	—	—	L	H	1–2 weeks	1-month storage at –0.5°C, 10% O_2, and 2–3% CO_2
tamarillos, tree tomatoes	*Cyphomamdra betacea*	3–5	37–41	90–95	—	—	VL	M	5–9 weeks	—
tamarind	*Tamarindus indica*	2–7	36–45	90–95	–3.7	25.4	VL	VL	3–4 weeks	—
taro, malangas, cocoyams, dasheens, eddoes	*Colocasia esculenta*	7–10	45–50	80–95	–0.9	30.3	VL	L	18 weeks	—
tomatillos, husk tomatoes	*Physalis ixocarpa*	5–10	41–50	80–90	—	—	M–L	M	3 weeks	chilling injury at 5°C for 3 weeks
tomatoes, firm-ripe	*Solanum lycopersicum*	13	55	85–90	–0.5	31.5	M	H	1–3 weeks	3–5% O_2 and 3–5% CO_2
tomatoes, mature-green	*Solanum lycopersicum*	13	55	90–95	–0.5	31.5	M	H	2–5 weeks	3–5% O_2 and 2–3% CO_2
truffles	*Tuber* spp.	0	32	90–95	—	—	VL	L	20–30 days	MAP beneficial
turnips	*Brassica campestris, Rapaceum group*	0	32	90–95	–1.0	30.1	VL	L	4–5 months	—
water chestnuts, matais	*Eleocharis dulcis*	0	32	98–100	—	—	—	—	10 months	store submerged in NaOCl solution
water convolvulus	*Ipomea aquatica*	12–14	54–57	90–95	—	—	M–L	H	10–12 days	—
watercress	*Nasturtium officinale*	0	32	>95	–0.3	31.5	VL	H	2–3 weeks	≥ 5% O_2 + >7% CO_2
watermelons	*Citrullus lanatus*	10–15	50–59	90	–0.4	31.3	L	M	2–3 weeks	no CA or MAP benefit
wax apples	*Syzygium samarangense*	12–14	54–57	90–95	—	—	VL	—	10–14 days	MAP reduces chilling and decay

Continued on next page

Appendix A-1: Storage recommendations for produce, continued

Common names	Scientific name	Storage temperature		Relative humidity (%)	Highest freezing temperature		Ethylene production	Ethylene sensitivity	Approximate storage life	Beneficial CA and MAP conditions, storage comments
		°C	°F		°C	°F				
yams	*Dioscorea* spp.	15	59	70–80	–1.1	30.0	VL	L	2–7 months	—
yucca	See cassava									

KEY

CA = controlled atmosphere

MAP = modified-atmosphere packaging

Ethylene production rates:

- VL = very low: <0.1 microliters per kilogram-hour (ul/kg-hr) at typical storage temperatures
- L = low: 0.1–1.0 ul/kg-hr
- M = moderate: 1.0–10 ul/kg-hr
- H = high: 10–100 ul/kg-hr
- VH = very high: >100 ul/kg-hr

Ethylene sensitivity level (detrimental effects include yellowing, softening, increased decay, abscission, browning):

- L = low: >10 ul/kg-hr
- M = moderate: 1–10 ul/kg-hr
- H = high: <1 ul/kg-hr

Sources:

Brecht, J. K., S. A. Sargent, P. E. Brecht, J. Saenz, and L. Rodowick. 2019. Protecting perishable foods during transport by truck or rail. U.S. Department of Agriculture Agricultural Marketing Service Handbook 669 and University of Florida Institute of Food and Agricultural Sciences HS1328. https://edis.ifas.ufl.edu/hs1328

Cantwell, M. I. 2002. Appendix: Summary table of optimal handling conditions for fresh produce. In Kader, A. A., ed., Postharvest technology of horticultural crops. Oakland: UC Agriculture and Natural Resources Publication 3311.

Reid, M. S. 2016. Cut flowers and greens. In K. C. Gross, C. Y. Wang, and M. E. Saltveit, eds., The commercial storage of fruits, vegetables, and nursery stocks. U.S. Department of Agriculture Agricultural Research Service Agriculture Handbook 66. https://www.ars.usda.gov/arsuserfiles/oc/np/commercialstorage/commercialstorage.pdf

Thompson, A. K. 2014. Fruit and vegetables: Harvesting, handling, and storage. 3rd ed. Hoboken, NJ: Wiley Blackwell.

Whiteman, T. M. 1957. Freezing points of fruits, vegetables, and florist stocks. Marketing Research Report 310551. U.S. Department of Agriculture, Agricultural Marketing Service, Transportation and Marketing Program. https://ideas.repec.org/p/ags/uamsmr/310551.html

Note: "—" in table cells indicates that data are not available or that certain table cells do not apply to certain harvest products.

Appendix A-2: Storage recommendations for cut flowers and greens

Common names	Scientific name	Storage temperature		Ethylene sensitivity	Approximate storage life	Storage comments
		°C	°F			
alstroemeria	*Alstroemeria cultivars*	0–1	32–34	S	long	stem bending with poor temperature control
anemone, windflower	*Anemone* spp.	0–1	32–34	S	short	Store vertically
anthurium, flamingo flower	*Anthurium andraeanum*	12.5–20	55–68	NS	3 weeks	2% O_2 beneficial
asparagus fern, lace fern, sprengeri	*Asperagus* spp.	0–1	32–34	S	—	—
aster, Michaelmas daisy	*Aster* spp.	0–1	32–34	NS	—	—
baby's breath, gypsophila	*Gypsophila paniculata*	0–1	32–34	S	3 weeks	in preservative solution
bird of paradise	*Strelitzia reginae*	6–7	43–45	NS	4 weeks	with fungicide at 7°C (45°F) and 85–90% relative humidity
bouvardia	*Bouvardia* spp.	0–1	32–34	S	—	—
calla lily	*Zanteschia* spp.	0–1	32–34	NS	—	—
carnation, mini carnation	*Dianthus caryophyllus*	0–1	32–34	S	2–5 weeks	2–4 weeks for open flowers; 4–5 weeks for closed flowers
chrysanthemum, florist mum	*Dendranthema x grandiflorum*	0–1	32–34	NS	2–4 weeks	1.2 centimeter buds for 2 weeks; 1.6 centimeter buds for 3 weeks; full mature blooms stored dry at 0°C for 3–4 weeks
daffodil	*Narcissus cultivars, Daffodils*	0–1	32–34	S	2 weeks	Store vertically at 90% relative humidity; 100% N_2 allows several weeks of storage and long vase life
delphinium, larkspur	*Delphinum, Consolida* spp.	0–1	32–34	S	—	—
emerald palm	*Chamaedorea* spp.	12.5	55	NS	1–2 weeks	—
eucalyptus, silver dollar tree	*Eucalyoptus* spp.	0–1	32–34	NS	—	Poorly handled foliage can produce ethylene
fir, spruce, pine	*Abies, picea, Pinus* spp.	0–1	32–34	NS	—	Wrap in plastic to reduce water loss
freesia	*Freesia x cultivars*	0–1	32–34	S	—	Young buds fail to open with ethylene exposure
gerbera, transvaal daisy	*Gerbera jamesonii*	0–1	32–34	NS	1 week	Stem bending reduced at proper storage temperature

Continued on next page

Appendix A–2: Storage recommendations for cut flowers and greens, continued

Common names	Scientific name	Storage temperature		Ethylene sensitivity	Approximate storage life	Storage comments
		°C	°F			
ginger, shell ginger, torch ginger	*Alpina zerumbet, A. purpurata*	12.5–15	55–59	NS	—	—
gladiolus	*Gladiolus cultivars*	0–1	32–34	S	1 week	Store upright
heliconia, parrot flower	*Heliconia humilis, H. psittacorum*	10–12.5	50–55	NS	2 weeks	—
holly, ilex	*Ilex* spp.	0–1	32–34	S	—	Some holly types are not ethylene-sensitive
huckleberry	*Vaccinium ovatum*	0–1	32–34	NS	—	—
iris, fleur-de-lis	*Iria* cvs.	0	32	NS	—	Store dry and upright
leatherleaf fern	*Rumohra adiantiformis*	1–6	34–43	NS	—	—
lemonleaf, salal	*Gaultheria shallon*	–0.5 to 1	31–34	NS	—	—
liatris, gay feather	*Liatrus pycnostachya, L. spicata*	0–1	32–34	NS	—	—
lily, Asiatic lily, Oriental lily	*Lilium* spp.	0–12.5	32–55	S	4 weeks	With silver thiosulfate or 1-methylcyclopropene, fungicide and sugar pulse treatments, storage temperature depends on cultivar
lily of the Nile, agapanthus	*Agapanthus africanus*	0–1	32–34	S	—	—
lisianthus, prairie gentian		0–1	32–34	S		Store upright if handled above recommended temperature
marguerite daisy, Boston daisy	*Argyranthemum frutescens*	0–1	32–34	NS	3 days–1 week	3 days in water and 1 week dry storage
orchids	*Cattleya, Cymbidium* cvs., *Dendrobium, Phalaenopsis, Vanda, Paphiopedium*	0–12.5	32–55	S	long	*Cymbidiums* have 4-week storage life; storage temperature depends on cultivar
protea, pincushion	*Proteacea* family	0–1	32–34	NS	—	—
rose, spray rose, sweetheart rose	*Rosa* cultivars, hybrids	0–1	32–34	S, some cultivars	2 weeks	In dry storage near 0°C (32°F)

Continued on next page

Appendix A–2: Storage recommendations for cut flowers and greens, continued

Common names	Scientific name	Storage temperature		Ethylene sensitivity	Approximate storage life	Storage comments
		°C	°F			
snapdragon	*Antiurrhinum majus* cultivars	0–1	32–34	S	7–10 days	Some new varieties are less ethylene-sensitive; wrap in plastic to retard moisture loss; store upright
statice, German statice	*Limonium* spp., *Goniolimon*	0–1	32–34	S	—	—
stock	*Matthiola incana*	0–1	32–34	S	—	—
sunflower	*Helianthus annuus*	0–1	32–34	S	—	stem bending with poor temperature control
sweet pea	*Lathyrus odoratus*	0–1	32–34	S	1 week	with 1-methylcyclopropene and sucrose treatment
sweet william	*Dianthus barbarous*	2–3	36–37	S	—	—
tuberose	*Polanthes tuberosa*	0–1	32–34	S	6 days	with sucrose treatment; slightly sensitive to ethylene, usually not treated
tulip	*Tulip* cultivars, hybrids	0–1	32–34	NS	variable by cultivar	Store dry and upright to prevent stem bending, 85% relative humidity
waxflower, geraldton waxflower	*Chamelaucium unicinatum*	0–1	32–34	S	2 weeks	—
yellow aster	*x Solidaste luteus*	0–1	32–34	NS	—	—

KEY

S = sensitive to ethylene

NS = not sensitive to ethylene

Source: Gross, K. C., C. Y. Wang, and M. E. Saltveit. 2016. The commercial storage of fruits, vegetables and nursery stocks. U.S. Department of Agriculture Agricultural Research Service Agricultural Handbook 66. https://www.ars.usda.gov/northeast-area/beltsville-md-barc/beltsville-agricultural-research-center/food-quality-laboratory/docs/ah66/

Note: "—" in table cells indicates that data are not available or that certain table cells do not apply to certain harvest products.

Appendix B: Cooling Times

Appendix B–1: Room cooling times

Commodity	Container, packing method, stacking pattern, air circulation (reference)	Seven-eighths cooling*	
		Average	Slowest
apples	wooden box, unpacked, pallet loads (Sainsbury 1961)	—	2–3.3 days
	wooden box, packed, pallet loads (Sainsbury 1961)	—	6–8 days
artichokes	corrugated container (unpublished test data)	24 hr	—
grapes	wooden lug, solid stacked, on pallets (Guillou 1960)	—	30 hr
nectarines	Plastic tray pack, packed and stacked, on pallets, room air speed 400 feet per minute (Mitchell et al. 1971)		
	wooden container, 20.5% side vent area (unpublished test data)	—	12 hr
	corrugated container, 10–13% side vent area (unpublished test data)	—	12 hr
	corrugated container, 6% side vent area (unpublished test data)	—	12 hr
	corrugated container, 1.9–3.4% side vent area (unpublished test data)	—	27 hr
oranges	Fruit in 47-in. square wood pallet bins (unpublished test data)		
	24 in. deep, no side vents	33 hr	—
	30 in. deep, no side vents	45 hr	—
pears	wooden box, 48 lb, paper-wrapped and unwrapped fruit, on pallet (Perry and Perkins 1968; Sainsbury 1961)	2.5 days	3–4 days
	LA lug, 24 lb (unpublished test data)	16 hr	20 hr
	LA lug, corrugated, 24 lb, paper-wrapped fruit (unpublished test data)	30 hr	39 hr
	Corrugated telescope container, 12 by 18 by 10 in. high, 36 lb, tight fill, boxes cross-stacked on pallet (Guillou 1960)		
	54 containers per pallet, two 1 1⁄4-in. holes in each side, 1⁄2-in. space between sides	2.5 days	4–6 days
	48 containers per pallet, two 1 1⁄4-in. holes in each side, 3 3⁄8-in. space between sides	2 days	3.6 days
	3.5% side vent, 1-in. spacing between containers	16 hr	23 hr
	5% side vent, no space between containers	24 hr	40 hr
plums	Corrugated containers, 11 by 17 1⁄2 by 8 in. high, 28 lb, tight-fill, cross-stacked on pallet (unpublished test data)		
	no vent and no spacing between containers, cross-stacked	—	84 hr
	4% vent area, 1-in spacing between containers, cross-stacked	—	22 hr
	4% vent area, no spacing, register-stacked	—	18 hr
	single 4-basket crate, 2 sides exposed	—	27 hr

Continued on next page

Appendix B–1: Room cooling times, continued

Commodity	Container, packing method, stacking pattern, air circulation (reference)	Seven-eighths cooling*	
		Average	Slowest
plums	Fruit in 47-in. square wood pallet bins (unpublished test data)		
	24 in. deep, no side vents	—	24–46 hr
	24 in. deep, six 5⁄8 by 16-in. vents on each side	—	23–40 hr
	metal bin, perforated, 1⁄4-in. holes on 1-in. centers	—	18–33 hr
	metal bin, perforated, six 5⁄8 by 16-in. slots on each side	—	20–32 hr

Sources:

Guillou, R. 1960. Coolers for fruits and vegetables. Bulletin of the California Agricultural Experiment Station 773.

Mitchell, F. G., R. A. Parsons, and G. Mayer. 1971. Cooling trials with plastic tray pack nectarines in various containers. California Agriculture 25(9):13–15. http://calag.ucanr.edu/Archive/?article=ca.v025n09p13

Perry, R. L., and R. M. Perkins. 1968. Hydrocooling sweet corn. St. Joseph, MI: American Society of Agricultural Engineers Paper 68–800.

Sainsbury, G. F. 1961. Cooling apples and pears in storage rooms. U.S. Department of Agriculture Marketing Research Report 474. https://doi.org/10.22004/ag.econ.312604

Note: "—" in table cells indicates that data are not available or that certain table cells do not apply to certain harvest products.

*Times are approximate and are to be used only as guides. Small differences in containers, packing procedures, or exposure to the coolant can affect cooling rates.

Appendix B–2: Hydrocooling times

Commodity	Postharvest handling or packing (reference)	Seven-eighths cooling times*	
		Average	Slowest
asparagus	bunches immersed in flowing water (Pentzer et al. 1936)	6 min	—
cantaloupes	bulk in commercial hydrocooler (Lipton and Stewart 1959)	45–60 min	—
peaches	47-in. square wood bins, 24 in. deep, water shower 150 gallons per minute per bin (O'Brien and Gentry 1967)		
	bin bottom with slots allowing free drainage	30 min	—
	bin bottom with ten 1-in. holes allowing bin to fill	24 min	—
	commercial hydrocooler (Touissant et al. 1955)	33 min	—
pears, Anjou	open lug, 150 fruit, water spray 4 gallons per minute per square foot (unpublished test data)	42 min	—
sweet corn	Wirebound wood crates (Perry and Perkins 1968)		
	immersion, no agitation	46–84 min	—
	immersion, agitation	28 min	—
	wirebound crates, water shower 5 gallons per minute per square foot, free drainage (unpublished test data)	45 min	—

Sources:

Lipton, W. J., and J. K. Stewart. 1959. Commercial cooling of cantaloupes tested. Western Grower and Shipper 30(6).

O'Brien, M., and J. P. Gentry. 1967. Effect of cooling methods on cooling rates and accompanying desiccation of fruits. Transactions of the ASAE 10(5):603–606. https://doi.org/10.13031/2013.39740

Pentzer, W. T., R. L. Perry, G. C. Hanna, J. S. Wiant, and C. E. Asbury. 1936. Precooling and shipping California asparagus. UC Agricultural Experiment Station Bulletin 600.

Perry, R. L., and R. M. Perkins. 1968. Hydrocooling sweet corn. St. Joseph, MI: American Society of Agricultural Engineers Paper 68–800.

Toussaint, W. D., T. T. Hatton, and G. Abshier. 1955. Hydrocooling peaches in the North Carolina sandhills. North Carolina Agricultural Experiment Station Agric. Env. Infor. Ser. 320.

Note: "—" in table cells indicates that data are not available.

*Times are approximate and are to be used only as guides. Small differences in containers, packing procedures, or exposure to the coolant can affect cooling rates.

Appendix B–3: Airflow rates, static pressure, and cooling times for forced-air cooling common produce

Commodity and container, containers placed three deep in register in cooling	Airflow rate [cubic feet per minute per pound] (static pressure [in. of water column])* Seven-eighths cooling time for downstream produce (hr)							
	1.5	2	3	4	6	9	12	18
artichokes, corrugated container, 9% side vent area	—	—	1.5 (2.4)	1.0 (1.2)	—	—	—	—
cantaloupes, corrugated containers, 6% side vent area	—	4 (NA)	2.0 (2.0)	1.2 (1.0)	0.6 (0.4)	0.3 (0.15)	0.2 (0.02)	—
cherries, 18 lb wood box with top cleat	—	—	—	0.9 (0.20)	0.5 (0.07)	0.3 (0.15)	0.0 (0.02)	—
grapes, TKV container (wood ends and paper-wood laminated sides, lid, and bottom), 3–5% side vent area, no bottom cleat	—	—	2.8 (>4)	1.6 (2–3)	0.7 (0.4–0.7)	0.4 (0.2)	0.25 (0.1)	—
grapes, TKV container (wood ends and paper-wood laminated sides, lid, and bottom), 3–5% side vent area, 3/4 in bottom cleat for added separation between containers	—	—	—	2.5 (0.7)	1.3 (0.25)	0.7 (0.1)	0.5 (0.05)	—
grapes, partial telescoping corrugated containers, 17.25 in. by 15 in. by 6 in. high, 5.8% side vent area	1.4 (3.0)	1.0 (1.7)	0.6 (0.7)	0.4 (0.4)	0.27 (0.16)	0.17 (0.07)	—	—
grapes, partial telescoping corrugated containers, 17.25 in. by 14 in. by 6.75 in. high, 3.5% side vent area	1.5 (>4)	1.1 (2.6)	0.7 (1.1)	0.5 (0.5)	0.3 (0.23)	0.22 (0.9)	—	—
kiwifruit, corrugated flats, single-layer tray, polyethylene film box liner (no venting), no top pad	—	—	—	—	0.8 (1.3)	0.5 (0.8)	0.26 (0.4)	—
kiwifruit, corrugated flats, single-layer tray, polyethylene film box liner (no venting), with top pad above fruit	—	—	—	—	1.5 (NA)	0.9 (2.1)	0.4 (0.4)	—
nectarines (or peaches), corrugated containers with plastic trays, 6% side vent area	NA (0.9)	NA (0.4)	0.8 (0.15)	0.5 (0.4)	0.20 (<0.01)	0.1 (<0.01)	—	—
nectarines, Bliss, corrugated container, 2-layer tray, top pad, 5% side vent area, standard pack	—	—	—	0.8 (0.9)	0.5 (0.4)	0.3 (0.1)	0.2 (0.06)	—
nectarines, Bliss, corrugated container, 2-layer tray, top pad, 5% side vent area, poly wrap top tray	—	—	—	1.0 (1.7)	0.6 (0.7)	0.4 (0.15)	0.25 (0.08)	—
peaches (or nectarines), plastic trays in wood lugs, 20% side vent area	NA (2.3)	1.6 (0.7)	1.0 (0.3)	0.7 (0.1)	0.4 (0.03)	0.25 (<0.01)	—	—
pears, corrugated container, 12 in. by 18 in. by 9.5 in. high, 2% side vent area	—	—	1.2 (NA)	0.8 (4)	0.5 (1.5)	0.3 (0.5)	0.2 (.25)	—

Continued on next page

Appendix B-3: Airflow rates, static pressure, and cooling times for forced-air cooling common produce, continued

Commodity and container, containers placed three deep in register in cooling	Airflow rate [cubic feet per minute per pound] (static pressure [in. of water column])* Seven-eighths cooling time for downstream produce (hr)							
	1.5	2	3	4	6	9	12	18
pears, corrugated container, 12 in. by 18 in. by 9.5 in. high, 5% side vent area	—	—	1.0 (1.8)	0.7 (0.6)	0.4 (0.2)	0.2 (0.09)	0.15 (<0.1)	—
pears, 46 lb corrugated container, paper-wrapped fruit, 5% side vent area	—	—	—	—	—	0.7 (2.1)	0.4 (1.1)	0.22 (0.5)
plums, corrugated container, 11 in. by 17.5 in. by 8 in. high, 4% side vent area	—	—	1.2 (2.9)	0.8 (1.3)	0.4 (0.3)	0.25 (0.15)	0.18 (NA)	—
strawberries, corrugated open-top tray, plastic mesh baskets	2.0 (0.4)	1.4 (0.2)	0.8 (0.08)	0.5 (0.04)	0.3 (0.02)	0.2 (<0.01)	—	—
tomatoes, corrugated container, 10% side vent area	NA (2.3)	2.9 (1.2)	1.6 (0.8)	1.1 (0.4)	0.6 (0.2)	0.4 (0.13)	0.25 (<0.01)	—
oranges, 2 ft–deep wood bins, vertical airflow through slots in bottom	—	—	1.0 (0.09)	0.6 (0.04)	0.4 (0.02)	0.2 (<0.01)	0.15 (<0.01)	—
oranges, 3 ft–deep wood bins, vertical airflow through slots in bottom	—	—	1.0 (0.2)	0.6 (0.10)	0.4 (0.04)	0.2 (0.02)	0.15 (<0.01)	—
oranges, 4 ft–deep bins, vertical airflow through slots in bottom	—	—	0.9 (0.2)	0.6 (0.07)	0.4 (0.03)	0.3 (0.02)	0.15 (<0.01)	—
pears, wood bins 4 ft by 4 ft by 2 ft, horizontal airflow, 5.2% side vent area	—	—	—	—	0.7 (0.4)	0.5 (0.15)	0.3 (0.08)	0.2 (0.05)
plums, average fruit, wood bins 4 ft by 4 ft by 2 ft, horizontal airflow, 3% side vent area	—	—	NA (0.9)	0.7 (0.25)	0.4 (0.07)	0.2 (<0.01)	0.1 (<0.01)	—
plums, slowest-cooling fruit, wood bins 4 ft by 4 ft by 2 ft, horizontal airflow, 3% side vent area	—	—	—	—	1.0 (1.8)	0.6 (0.7)	0.4 (0.3)	0.2 (0.12)

Sources: Mitchell et al. 1971; Parsons, et al. 1970, 1972; Wang and Tupin 1968

Note: "—" in table cells indicates that data are not available or that certain table cells do not apply to certain harvest products.

*Static pressure (in parentheses) is for an air path through three tiers stacked in register. For other stacking arrangements, multiply the static pressure by 0.05 for one tier, 0.3 for two tiers, 2.3 for four tiers, and 7.0 for six tiers. Data are approximate only; small differences in container venting and stacking patterns can cause large changes in static pressure losses.

Appendix B–4: Airflow rates, static pressure, and cooling times for forced-air cooling of selected flowers in corrugated containers*

Flower	Box size (in.)	Vent hole diameter (in.)	Gross weight (lb)	Static pressure					
				0.5 in.		1.0 in.		2.0 in.	
				Airflow (cfm/ box)	Seven-eighths cooling time (min)	Airflow (cfm/ box)	Seven-eighths cooling time (min)	Airflow (cfm/ box)	Seven-eighths cooling time (min)
carnation	48 by 21 by 12	2	51	70	48	90	40	110	35
chrysanthemum	57 by 21 by 12	2	33†	80	62	130	58	210	54
gypsophilia	42 by 21 by 12	3	—	120	10	260	8	—	—
rose	48 by 21 by 12	2	20 bunches per box	140	34	200	25	290	20
statice	42 by 21 by 12	2	75	150	40	210	18	280	13

KEY

cfm = cubic feet per minute

Note: "—" in table cells indicates that data are not available or that certain table cells do not apply to certain harvest products.

*Two holes in each end panel. Air flows through the end panels along the length of a single box.

†45-lb box allows 50% less airflow through box.

Appendix C: Useful constants and conversions

Appendix C–1: Refrigeration effect conversions

Refrigeration source	Cooling effect
1 ton of refrigeration	Heat needed to melt water ice at a rate of 1 ton in 24 hr • 288,000 Btu/day • 12,000 Btu/hr • 200 Btu/min
melting 1 lb of water ice	144 Btu
melting 1 kg of water ice	335 kJ
vaporizing 1 lb of water	1,073 Btu at 32°F or 1,035 Btu at 100°F
vaporizing 1 kg of water	2,496 kJ at 0°C or 2,408 kJ at 38°C
vaporizing 1 lb of dry ice and warming vapor to 32°F	264 Btu (equal to melting 1.8 lb of water ice)
vaporizing 1 kg of dry ice and warming vapor to 0°C	614 kJ (equal to melting 1.8 kg of water ice)
vaporizing 1 lb of liquid nitrogen or 1 lb of liquid carbon dioxide and warming the vapor to 32°F	175 Btu (equal to melting 1.2 lb of water ice)
vaporizing 1 kg of liquid nitrogen or 1 kg of liquid carbon dioxide and warming the vapor to 0°C	407 kJ (equal to melting 1.2 kg of water ice)

Appendix C-2: Approximate compressor horsepower per ton of refrigeration

Evaporating temperature (°F)	Condensing temperature (°F)		
	95	115	135
30	1.1	1.5	2.0
20	1.3	1.8	2.4
10	1.5	2.1	2.7

Appendix C-3: Approximate compressor power in kW per kW of refrigeration

Evaporating temperature (°C)	Condensing temperature (°C)		
	35	45	55
0	0.23	0.29	0.39
−5	0.26	0.35	0.46
−10	0.30	0.41	0.51

Appendix C-4: Heat input to refrigerated spaces

Heat source	Heat input
1 shaft horsepower (fans or pumps with motor outside refrigerated space)	2,540 Btuh 0.21 tons of refrigeration 0.74 kW of refrigeration
1 kW (lights, motors)	3,413 Btuh 0.28 tons of refrigeration space 1 kW of refrigeration

Appendix C-5: Physical properties of materials

Property	Customary units	SI (metric) units
specific heat of fruit and vegetables	0.85–0.95 Btu/lb/°F, average 0.90 Btu/lb/°F	3.56–3.98, average 3.77 $kJ \cdot kg^{-1} \cdot °C^{-1}$
specific heat of dry wood and paper	0.30 Btu/lb/°F	1.26 $kJ \cdot kg^{-1} \cdot °C^{-1}$
specific volume of air at 0°C (32°F)	12.5 ft^3/lb	0.78 $m^3 \cdot kg^{-1}$
density of ice in blocks	56 lb/ft^3	897 $kg \cdot m^{-3}$
density of water at 0°C (32°F)	62.4 lb/ft^3, 8.3 lb/gal	1000 $kg \cdot m^{-3}$

Appendix C–6: Measurement conversions

U.S. customary	Conversion factor for U.S. customary to SI (metric)	Conversion factor for SI (metric) to U.S. customary	SI (metric)
Length, area, volume			
inch (in)	2.540	0.3937	centimeter (cm)
foot (ft) (=12 in)	0.3048	3.281	meter (m)
yard (yd) (=3 ft)	0.9144	1.094	meter (m)
mile (mi) (=5280 ft)	1.609	0.6215	kilometer (km)
acre (ac) (=43560 ft^2)	0.4047	2.471	hectare (ha)
square foot (ft^2)	0.0929	10.764	square meter (m^2)
square mile (mi^2)	2.590	0.3861	square kilometer (km^2)
fluid ounce (fl oz)	29.57	0.03381	milliliter (ml)
quart, liquid (qt)	0.9463	1.056	liter (l)
quart, dry (qt)	1.101	0.9083	liter (l)
bushel (bu)	0.03524	28.38	cubic meter (m^3)
gallon (gal)	3.785	0.2642	liter (l)
cubic foot (ft^3)	28.32	0.3531	liter (l)
cubic foot (ft^3)	0.02832	35.31	cubic meter (m^3)(=1000 l)
cubic yard (yd^3)	0.765	1.307	cubic meter (m^3)
Flow			
cubic feet per minute (ft^3/min)	0.000472	2118.6	cubic meters per second ($m^3 \cdot sec^{-1}$)
cubic feet per minute per pound (ft^3/min/lb)	0.00104	961.5	($m^3 \cdot sec^{-1} \cdot kg^{-1}$)
cubic feet per minute per pound (ft^3/min/lb)	1.04	0.9615	cubic meters per second per metric ton ($m^3 \cdot sec^{-1} \cdot t^{-1}$)
cubic feet per minute per pound (ft^3/min/lb)	1.04	0.9615	liters per second per kilogram ($l \cdot sec^{-1} \cdot kg^{-1}$)
gallons per minute (gpm)	0.0631	15.85	liters per sec ($l \cdot sec^{-1}$)
gallons per minute per square foot (gpm/ft^2)	0.6791	1.473	liters per sec per square meter ($l \cdot sec^{-1} \cdot m^2$)

Continued on next page

Appendix C–6: Measurement conversions, continued

U.S. customary	Conversion factor for U.S. customary to SI (metric)	Conversion factor for SI (metric) to U.S. customary	SI (metric)
Mass, pressure, density			
ounce (oz)	28.35	0.035	gram (g)
pound (lb)	0.4536	2.205	kilogram (kg)
ton (T)	0.9072	1.102	metric ton (t) (=1000 kg)
pound per square inch (psi)	6.895	0.1450	kilopascal (kPa)
pound per square inch (psi)	0.0689	14.51	bar
pound per square inch (psi)	51.713	0.01934	mmHg @ 0°C (=0.1333 kPa)
inches of water column (in w.c.)	0.2487	4.021	kilopascal (kPa)
pound per cubic foot (lb/ft^3)	16.02	0.06242	$kg{\cdot}m^{-3}$ (= $g{\cdot}l^{-1}$)
Energy, power, heat			
BTU	1.055	0.9479	kilojoule (kJ) (=kW·sec)
horsepower (hp)	0.7457	1.341	kilowatt (kW)
BTU per pound per °F	4.1868	0.2388	$kJ{\cdot}kg^{-1}{\cdot}C^{-1}$
ton of refrigeration	3.519	0.2842	kW of refrigeration
Temperature			
Fahrenheit (°F)	°C = (°F – 32) ÷ 1.8	°F = (°C x 1.8) + 32	Celsius (°C)

Index

Page numbers with an italic *t*, such as 8*t*, refer to tables.

Printed in the USA
CPSIA information can be obtained
at www.ICGtesting.com
JSHW072135100124
55197JS00007B/12

9 781627 110594